PROJECT PORTFOLIO MANAGEMENT

PROJECT PORTFOLIO MANAGEMENT
A VIEW FROM THE
MANAGEMENT TRENCHES

The Enterprise Portfolio Management Council

WILEY

John Wiley & Sons, Inc.

Project Management Institute

For general information about our other products and services, please contact our Customer Care Department within the United States at (800) 762-2974, outside the United States at (317) 572-3993 or fax (317) 572-4002.

Wiley also publishes its books in a variety of electronic formats. Some content that appears in print may not be available in electronic books. For more information about Wiley products, visit our web site at www.wiley.com.

Project Management Institute (PMI) is the world's leading project management organization. PMI shares knowledge with nearly 500,000 members and credential holders in 185 countries. Since its founding 40 years ago, PMI has impacted more than one million practitioners, businesses, governments, students, and training organizations. Today, PMI's products and services range from world-class standards for project, program, and portfolio management to five professional credentials, including the Project Management Professional (PMP)®. Learn more at www.PMI.org.

Library of Congress Cataloging-in-Publication Data:

Project portfolio management: a view from the management trenches / The Enterprise Portfolio Management Council.
    p. cm.
  Includes bibliographical references and index.
  ISBN: 978-0-470-50536-6 (cloth)
  1. Project management.   I. Enterprise Portfolio Management Council.
  HD69.P75P654 2009
  658.4′04–dc22

                                                            2009019342

Printed in the United States of America

10 9 8 7 6 5 4 3 2 1

Contents

Foreword

"Not to decide is to decide."

—Dr. Harvey Cox, *The Secular City*

This book, like much of the PPM work produced by Gartner Inc. over the past several years, helps take the "not" out of "not to decide." Following the guidelines the authors have clearly set forth, companies can take the initiative in obtaining clear decision criteria from management about which projects to do. The book shows how a company can put in place intentional procedures to sort through the myriad project proposals to help discern those that promise the most value to the company. In addition, the book shows how a company can optimally allocate resources to work by aligning ITs efforts and keeping them aligned. In short, it illuminates how the discipline known as PPM can help overcome the syndrome whereby random processes deliver random results to reliably add value to the company's bottom line.

You may be familiar with the usual process occurring in many companies now: a department head submits a project proposal and after a few weeks or months and after a few reminders, the project works its way to the head of the line and IT management allocates a team to it. A few months may go by before the project deliverable arrives. It may become obvious at that point that it is a duplicate to a system already in use in another department. You may then realize that this was the tenth project in three months for that department and that it may or may not support a declared company objective. In all likelihood it may have consumed scarce developer resources, forcing other projects to wait. Sound familiar?

"Not to decide is to decide." When companies don't have clear decision criteria, or defined processes to choose among proposed projects, or staff specifically allocated to map projects-to-goals, then it is no wonder that it is impossible to keep track of how much effort is spent in support of the various goals throughout the various departments. This book will enable management to sort through all the demand and to initiate and complete projects in an efficient and organized way. In this way, companies may avoid the tried-but-ineffective methods of addressing projects either in turn, or by coin toss, or based on good collegial relationships, or on a requester's political clout.

Matt Light
Research Vice President
Gartner, Inc.

"There are no secrets to success. It is the result of preparation, hard work, learning from failure."

—Colin Powell

Project Portfolio Management: A View from the Management Trenches will be a valuable addition to the toolkit of all portfolio managers. I really enjoyed the folksy, down to earth discussion of project portfolio management in this book. There is much wisdom shared by a diverse group of accomplished practitioners from multiple industries. The view from the trenches tells a compelling story of how the project portfolio management is actually practiced in multiple settings versus how things ought to be. My own early experiences and difficulties in implementing portfolio management as a VP of R&D made me a champion of portfolio management during my years as VP of Conferences and then President of the Product Development and Management Association (PDMA).

I realized I had achieved less-than-stellar success with project portfolio management at Rich-SeaPak in the 1990s. We did not have a clear mechanism to kill projects; as a result we had too many projects in the pipeline and no one was charged to prune the project portfolio. Decisions about projects were routinely made at a project level and not at the portfolio level. So, I missed a great opportunity to formulate and operationalize what this book refers to as strategic portfolio management at the Front End.

This book will help the reader avoid serious mistakes commonly made in the design and implementation of enterprise portfolio management systems in organizations. I wish the readers much success as they digest keys messages "from the trenches" and apply those insights to achieve stellar success in the practice of portfolio management.

Dr. Hamsa Thota
Former Vice President-Development, Rich-Sea Pak Corporation

Former Chairman and President, Product Development and Management Association

Director, New Product Institute

President, Innovation Business Development, Inc.

"Discontent is the first step in the progress of a man or a nation."

—Oscar Wilde

Or an IT strategy. Any executive who has confronted the perpetual challenge of business-IT alignment knows all too well how challenging it can be to determine which IT projects merit how much investment, or how a given project may fit in with the company's overall goals and priorities. From the data center to the business units, and certainly throughout the C-suite, discontent abounds.

So it's heartening, really, to think of that collective discontent as an essential first step toward progress. The question then becomes what constitutes the second step? This book holds a very important part of the answer. Among its virtues is that it provides a critical (and practical) context for IT decision-making that's relevant not only for those who will actually tackle IT projects, but for all the executives who play important roles in vetting and approving both specific projects and overall IT strategy.

That latter group certainly includes CFOs and other senior finance executives, who often struggle to find common ground or a common language in their discussions with IT leaders. The concept of portfolio management, however, is something many if not most of them are very familiar with. Borrowed from the investment world, it's a widely applicable technique that provides badly needed context for decision-making.

That context is nicely captured in a handful of key questions: is the company investing in the right things? How well does it execute? Can it absorb the changes certain projects will bring about? And so on. Those are not esoteric questions that require technological expertise. They are the foundation on which a company's IT strategy will be built. And they are questions that all key stakeholders within a company should address together.

Project portfolio management is not new, not even as it pertains to IT. To implement it successfully a company will probably have to make a number of changes across the organization. This book will demonstrate why that effort will pay off, and just what to expect along the way. If you know the feeling of discontent that surrounds IT project management, take that second step toward progress: read this book.

Scott Leibs

Executive Editor

CFO Magazine

"The definition of insanity is doing the same thing over and over and expecting different results."

—Albert Einstein

The challenge of realizing real business value from today's increasingly large and complex IT investments continues to be one for all enterprises. The success rate of IT projects still hovers around 30%—basically unchanged over the last 10-20 years that we have been measuring it. IT governance is gaining increasing attention among executives, business and IT management, IT professionals and IT auditors as a "solution" to this challenge. In *The Information Paradox*,[1] and in my work with ISACA and the IT Governance Institute (ITGI), leading the development of their Val IT[TM] Framework,[2] we identified portfolio management as a key element of effective governance. Portfolio management is a powerful tool to support decision making around anything that is important to an enterprise, including, but certainly not limited to IT projects. The combination of portfolio management and performance management provides the "engine" of effective governance.

Any engine needs fuel—the "governance engine" needs high-grade business cases. The seeds of success or failure are sown in the business case. As the authors state, the business case is a foundational, and essential tool for portfolio management. However, enterprises generally are not good at developing or using complete and comprehensive business cases. The current approach to business cases pretty well guarantees significant challenges, if not outright failure. A well developed and intelligently used business case is one of the most valuable tools available to management. The quality of the business case and the processes

---

1. Enterprise Value: Governance of IT Investments: The Val IT Framework 2.0, IT Governance Institute, 2008

2. The Information Paradox—Realizing the Business Benefits of Information Technology, written by John Thorp jointly with Fujitsu, first published in 1998 and revised in 2003, McGraw-Hill, Canada

involved in its creation and use—as a living operational document—have an enormous impact on creating and sustaining value.

Many books and articles have been written on the topics of governance and portfolio management yet, as the authors of this book point out, adoption continues to proceed at a snail's pace. One objection raised is that the concepts are too academic or theoretical. Where there is recognition of the need to act, the next questions are: "Where do we start?" "How do we do it?" and "How will we know we are successful?" The members of EPMC who contributed to this book are practitioners who collectively have many years experience in using effective business cases and portfolio management to improve their overall governance of IT, and to create and sustain real business value for their enterprises - these guys have been there and live this every day. This book provides valuable practical guidance for all who want to make a difference and help their enterprise realize the full potential of IT-enabled change.

John Thorp

Thought Leader,

Information Systems Audit and Control Association

IT Governance Institute

Preface

Effective project portfolio management has become a significant factor in the long-term strategic success of organizations. Its growth as a management practice can be attributed to its link to business policy and organizational strategy. The concept of PPM is fairly simple—you need to direct the right resources to efficiently deliver the right project investments in order to meet your organization's strategic goals and objectives. At the same time you need to deliver the quality and benefits that were expected before the work began.

Actually practicing PPM—and actually doing it well—is another story. And that's why we've brought together some of the leading practitioners of PPM to show you how it's being implemented effectively in a variety of organizations. This book is written by folks in the trenches, folks who successfully oversee the delivery of their organization's most critical programs and projects, folks who daily navigate their organization's political and cultural waters and effectively manage their stakeholders' expectations. These folks realize that PPM is a hard sell in some organizations, or that it's been attempted in some organizations but doesn't seem to gain the traction needed to be successful. And in some cases, organizations are sure where to start on what to do to make it work (knowing that you need to quickly demonstrate that PPM will bring value to the organization). It's for those folks (and maybe even more importantly, the executive stakeholders in the organization) that this book has been written.

So what is this "project portfolio management" concept? First, a portfolio is a range of investments. Synonyms include "collection," "aggregation," "variety," "accumulation," "multitude," "assortment," "ensemble." Therefore, a project portfolio is a collection of projects that, in the aggregate, make up an organization's investment strategy.

Project portfolio management is, therefore, the "centralized management of one or more portfolios, which includes identifying, prioritizing, authorizing, managing, and controlling projects, programs, and other related work to achieve specific strategic business objectives."[1] It seeks to answer the questions, What should we take on? and What should we drop? It requires achieving a delicate balancing of strategic and tactical requirements. PPM often requires determining what is possible (Do we have the capability, the resources?) and what is needed (Does it make good business sense?). Balancing capability and need generally results in defining the best that can be achieved with the limited resources available, rather than attempting to find the perfect solution (which in a perfect world would include infinite resources). When used effectively, PPM ensures that projects are aligned with corporate strategies and priorities and optimizes resource allocation. It's the practice that bridges the gap between the executive decision process and project execution.

Those of us who have already bought into PPM realize that effectively applying these practices has become increasingly critical to our businesses. All of our organizations, large and small, must select and manage their investments and execute their projects wisely to reap the maximum benefits from their investment decisions. A few of the most important benefits that have been realized include the following:

PPM enables our businesses to:

- Provide a structure for selecting the right projects and eliminating wrong ones
- Allocate resources to the right projects, thus reducing wasteful spending
- Align portfolio decisions to strategic business goals
- Base portfolio decisions on logic, reasoning, and objectivity
- Create ownership among staff by involvement at the right levels
- Establish avenues for individuals to identify opportunities and obtain support
- Help project teams understand the value of their contributions.

---

1. Project Management Institute (2008). *The Standard for Portfolio Management*—Second Edition, Newtown Square, PA: Project Management Institute.

So how well is your organization managing its portfolio of projects? You're answers to the following questions should offer you some insight.

- Does your portfolio reflect and support your business strategy?
- Is each project consistent with business strategy?
- Does the breakdown of project spending reflect your strategic priorities?
- Is the economic value of the total portfolio higher than what you've spent on it?
- Once projects start, what is the chance they'll ever be killed?
- Are projects being done in a time-efficient manner?
- Are project success rates and profit performance results consistent with expectations?
- Is the project portfolio heavily weighted to low-value, trivial, small projects?
- How are opinions of senior people and key decision-makers in the business captured in order to make project decisions?
- Have you considered what the right balance of projects for the new product portfolio is?
- Are there redundant projects being performed?
- Have all the projects in play been justified on solid business criteria?
- And of those that were approved, are they still justified?
- Do the managers and team members know where the projects they are working on fit into the priority ranking that best supports the business?
- Are there enough resources to get the work done; and if there are not, what trade-offs need to be made?
- Which projects make the most money?
- Which have the lowest risk?
- Which have subjective value, in terms of community image or internal morale?
- Which are not optional—projects dictated by regulatory requirements, for example?

We chose to present the first section of the book in narrative form to help you identify with our lead character who has just begun his PPM adventure. His story is not unusual—in fact, each of our authors went through similar growing pains. The story is interspersed with specific recommendations from the authors for how he can deal with the problems

that arise. The second half of the book shows you, in detail, what you need
to consider in implementing project portfolio management in your organization. We hope you'll see that the challenges confront you can be
overcome, and have been overcome by the authors.

Good reading,

James S. Pennypacker

Director, Center for Business Practices

Why We Created the Enterprise Portfolio Management Council

From the earliest days of the Enterprise Portfolio Management Council (EPMC), our goal was to provide pragmatic responses to the pervasive question in our profession: "If project portfolio management is so self-evident, why is there such limited traction; and, more importantly, what do you do about it?" We wanted to ensure that our perspective would have universal appeal. Our approaches to exploring the old issue of balancing growth with change in the face of limited resources should be read not only by portfolio, program, or project managers, but also by CEOs, CFOs, and other C-Level executives who care about this issue and about the real people and real dollars involved in their business' success.

In March, 2005, long before we had the idea for a book, a small group of likeminded project/portfolio practitioners met in San Francisco to form what has become the EPMC. Our belief was that there was a major disconnect between those who were making *decisions* on which projects to fund and those who were *tasked* with executing those projects.

We envisioned an organization that was free from consultants and vendors, where non-competing practitioners could share ideas, intellectual capital, software information, processes, problems, challenges, solutions, and future plans. It was from this vision that the EPMC was formed. As the founders of the EPMC, we suspected that there was an opportunity to accelerate the development, sophistication, and pragmatic application of enterprise portfolio management (EPM) methods resulting in greater business value for organizations through EPM. To that end, we proposed making the EPMC a *not-for-profit* organization.

The EPMC was chartered soon after our first meeting. The EPMC then took on the tasks of developing, deploying, and generating awareness of Enterprise Portfolio Management and the advancement of the tools, techniques, processes, and implementation methods that support it.

The EPMC grew to include experienced EPM practitioners representing major corporations and organizations across all industries, along with a smaller number of industry analysts and academics.

The format of the EPMC is that of peers representing a wide range of non-competitive industries and organizations, enabling the group to facilitate a more rapid expansion of the EPM body of knowledge and experience. One of the goals was to create a forum and a defined mechanism where member organizations could exchange best practices, assets, and processes. We believed that within this range of organizations, developed "assets", or Intellectual Property (IP), might be shared and exchanged without needing to make significant vendor software purchases.

A secondary goal of the Council was to grow industry awareness and understanding of the EPM method; including what it is, how it is implemented, tools and processes that support the method, and the challenges and advantages of a structured EPM approach. By educating other companies on the EPM approach, we helped bring new practitioners, new ideas, and new tools into the market and, over time, improved the quality of EPM deployment and corporate performance of our members' organizations.

The EPMC has continued to grow and support a wide range of activities, including: the creation of standards, certification; the dissemination of information on the EPM method and implementation; the creation of working groups around specific EPM topics; and the creation of an accepted, industry-wide EPM maturity model. Our reach is now global and has grown to include practitioners from adjunct fields.

It has been a long journey. When we started bringing these principles to light, the country was experiencing record profits on Wall Street and was in the midst of one of the longest growth periods in history. What a difference a couple of years make. While the principles outlined in the book apply to any economic period, they are especially critical in the radically changed business environment today.

We wrote this book to share our passion for the EPM philosophy and process with you, the reader. We offer our thanks to all of the current members and their respective organizations who made this book a reality. As we embarked on this daunting task, there were some of us who weren't sure we could pull it off. But with the same passion and structure we have for our EPM roles, we used laser focus, outcome orientation, and perseverance to write the volume you are about to read.

We hope you will take a moment to review each contributor's brief biography in order to get an overview of the broad appeal our principles have had in companies and organizations around the world.

We wish you all continued success and growth in your own Portfolio Management journey.

—San Retna and Mark Stabler

Why We Wrote the Book

As the Enterprise Portfolio Management Council members continued to learn more about project portfolio management (PPM) from each other over the years since its inception, we discovered a distinct shortage of practical books written to aid those of us practicing the art of PPM. Most of the volumes available on the topic of PPM spoke from an academic point of view, or from some proprietary source selling processes, methodologies, consulting services, and even software programs. Realizing that we were seeing distinct patterns of workable processes amongst ourselves, and encountering similar issues, we began to ask ourselves how we could best help others in our situation. We concluded that we could not be alone in this PPM journey. We bounced around a number of ideas, but the idea of writing a practical book excited the EPMC membership. We came to the conclusion that we could put our ideas into a format that could reach far more people, likely going through some of the same trials and tribulations as each of us in dealing with PPM. We decided to write a book about what we have learned in a clear and practical style that could communicate to many others some of the sound practices we have found to work, and to help avoid some of the difficulties we have learned about the hard way.

The next step, of course, was to ask for volunteers to actually determine what to specifically write about and then to actually complete a manuscript. In the end, ten of us decided it was something we indeed were willing to put the effort in to so a book could come to life. Starting with a blank sheet of paper is daunting for any writer. We decided to form subteams to do some legwork and prepare an outline for consideration. The task worked best using a technique called, among other things, a concept map. We converted the map into an outline with suggested topics. Next, of course, came the question of who would write which chapters. Again, volunteering worked to make the assignments. So, off we went . . . and soon had in hand several chapters in draft form. But, guess what, the styles were all different. Some of us wrote in an academic style, others wrote in a conversational style, and still others wrote in a narrative, or story, style. Now, which one should we choose? Good question!!

At this point we brought in an experienced editor for guidance and assistance. The group could not decide exactly which style to pursue, so the editor packaged up some samples from the three different styles and pulsed the publishing community. As it turned out, the package was quickly picked up by John Wiley & Sons, who sent it out to some independent assessors for review and comment. The result of the reviews told us that a conversational style and a narrative, or story, style were the most popular. We toyed with the idea of doing two books, but decided instead to write one book incorporating both styles!

The book starts with a brief introduction to the major concepts of PPM in a conversational style. The next six chapters explore the world of PPM through a story from the perspective of the central character, a relatively new portfolio manager. Through the course of his travails, the manager discovers some new concepts about PPM. The next five chapters of the book go back to a conversational style to expand on the points brought out about PPM in the story. The final chapter of the book catches up with our now wiser portfolio manager as he looks back over the past year and ponders what lies ahead.

We all enjoyed writing this book together and hope you gain a better understanding of PPM and pick up some helpful tips and guidance to use as you travel on your own PPM journey!

—The EPMC 10
Michael Gosnear
Stephen Jenner
Mike Mee
Michael M. Menke
Diane D. Miller
San Retna
Mark Stabler
Michael J. Stratton
Sarma Tekumalla
Mark Wybraniec

About the Authors

Michael Gosnear, MGM Mirage. Mike has over twenty-five years of information technology and operations experience, which includes government defense contracting, healthcare, and gaming industries. Over the past six years he has focused on portfolio management opportunities in healthcare and gaming, including developing governance, tools, and process methodologies across IT.

Stephen Jenner, UK Government. Steve has extensive experience of investment and benefits management in the public sector. The approach he developed to manage the criminal justice system IT portfolio was recognized internationally, in reports to the OECD and European Commission, and in a case study by Gartner. Steve is a professionally qualified management accountant with an MBA, and Masters of Studies from Cambridge University. He is also the author of *Realising Benefits from Government ICT Investments: A Fool's Errand?*

Michael Mee, Franklin Templeton Investments. Mike joined Franklin Templeton in 2004 as director of Franklin Templeton Technology Administration. In February 2005 he was promoted to vice president of Operations and Technology Administration. Mike facilitates the internal executive committee of the Operations and Technology Council (TOC). The role of the TOC is to approve, prioritize, and track technology projects within Franklin. Additionally, Mike is responsible for the continuing evaluation and improvement of the project portfolio management processes at Franklin Templeton.

Michael M. Menke, Value Creation Associates. Michael is the President of Value Creation Associates, a management consulting firm focused on the front end of the value chain: generating, evaluating and selecting the highest value opportunities for business and government. Formerly he was Fellow of Knowledge Management at Decision Strategies Inc., a strategy and decision consultancy. Prior to that he was Chief Portfolio Advocate at Hewlett-Packard. He has been helping senior leaders improve their strategies, decisions and portfolios since 1972 and helped introduce portfolio management as a best practice into the global pharmaceutical and oil & gas industries. He has taught executives about portfolio management and good decision making all over the world. He has a BA in physics from Princeton, an M.Sc. from Cambridge and a Ph.D. in physics from Stanford University.

Diane D. Miller, PMP, Independent Consultant (Former IT PMO Lead for Coca-Cola Enterprises, Inc.). Diane is a senior management professional with experience leading organizations through developing, implementing, and sustaining strong portfolio management practices. Her pragmatic approach to portfolio management is credited with saving millions of dollars through improved investment decisions and project delivery. Diane has earned the designation of PMP from the Project Management Institute and a BS in Management Information Systems from Syracuse University.

San Retna, Safeway. San is the vice president of IT Effectiveness and Optimization for Safeway, Inc. He has spent over twenty years in building great depth on pragmatically deploying and operating project, program, and portfolio management capabilities. Prior roles included managing principal at TransformAction and Chief Portfolio Officer at AAA of California, Utah, and Nevada. He also spent over ten years at Accenture, specializing in large-scale program management ($25M+). Clients included Bell Atlantic, Elis, JP Morgan, Merrill Lynch, Philip Morris, and Washington Mutual. His experience covers more than countries across four continents. This includes ten years in the United States, four years in the UK, and at least two years in France, Nigeria, and Switzerland. As a thought leader, he has been profiled in numerous publications including *CFO* magazine, *CIO* magazine, *Computerworld,* and *PM Network*. Case studies of successes have also been published by Gartner, Corporate Executive Board, META, and the Project Management Institute.

Mark Stabler, AAA for Utah, Nevada, and Northern California. Mark has nearly twenty years of program and project management experience garnered in leading corporations such as AAA, NCR, and AT&T. He has core expertise in developing and implementing processes across diverse business functions; project budgeting, tracking, and resource management; and project team management. He has helped create and deploy sophisticated Enterprise Portfolio Management infrastructures, enabling companies to consistently deliver value and realize return on internal investments. He speaks regularly at industry conferences, and is a founding member of the Enterprise Portfolio Management Council and serves on its board. He has published several white papers and articles.

Michael J. Stratton, The Boeing Company. Mike is a senior project manager at the Boeing Company in Washington State. His thirty-year career at Boeing includes working on nearly all of the commercial jetliners in a variety of positions including manufacturing engineering, make/buy, communications, and training. He holds PMP #611, a BA from Washington State University, an MBA from City University, and is pursuing his PhD focusing on project management at Capella University.

Sarma Tekumalla, Grange Insurance. Sarma has over twenty years experience in technology and project management. He managed local and global teams for start-up technology to large financial firms. He executed technology, process reengineering, and M&A projects to building an enterprise PMO from ground up. He is a Project Management Professional (PMP) and has his MBA from Fisher School of Business at Ohio State University.

Mark Wybraniec, Johnson & Johnson. Mark is a senior IT Executive with expertise in PMO and IT transformation. Mark has over eighteen years of experience as a senior IT executive in the strategic planning, analysis, and design of a variety of business information systems, across a broad range of technology platforms. He has substantial depth in successfully deploying and operating project, program, and portfolio management capabilities. Mark has a focus on how to best leverage the Program Management Organization function to drive greater value from project investments.

About the EPMC

OVERVIEW

The Enterprise Portfolio Management Council (EPMC) is a nonprofit organization whose goal is to support and accelerate the development of Enterprise Portfolio Management processes, tools, and techniques for the betterment of shareholder value. Portfolio management, though still in its early stages of maturity, can help corporations gain a competitive advantage by improving efficiencies, lowering costs, and increasing the return on internal investments. With hundreds of billions of dollars spent each year on internal projects and programs, portfolio management offers the potential for vast savings.

CHARTER

The EPMC provides an environment where experienced portfolio management practitioners can share ideas, technologies, processes, tools, challenges, and successes.

GOALS

The organization's goal is to create a community of senior portfolio management executives representing a broad spectrum of industries and organizations, in order to fuel the expansion of the portfolio management body of knowledge. Over time, the EPMC aims to create an open infrastructure that will help establish capability standards, and allow member organizations to exchange assets, such as processes and technologies. By educating companies on the portfolio management approach, the EPMC can help guide new practitioners, create new ideas and tools, and over time improve the quality of deployment and the performance of member organizations. To help achieve this goal, the EPMC has been featured in leading publications, including *CIO* magazine, CFO.com, and *Computerworld*.

Contact the EPMC at www.theepmc.org, or info@theepmc.org.

Acknowledgements

The journey through Portfolio Management has been very rewarding for me, as it has given me the opportunity to work for some of the best companies in the world and learn from some of the greatest business minds. Throughout this journey, I have been able to forge strong professional and personal relationships while implementing Portfolio Management at Johnson & Johnson and MGM Mirage. Along the way, several individuals have been instrumental in providing me with the guidance, opportunities, and insights I needed to excel in this field. I would also like to recognize and thank my fellow EPMC members who worked countless hours to make this book a reality. Lastly, I would like to thank my wife and family for all of their support and understanding, which have enabled me to pursue my passion for Portfolio Management and spend time away from them writing this book.

Michael Gosnear

I thank my family for their support and Franklin Templeton Investments for allowing me to learn, grow, and apply the profession.

Mike Mee

I owe a deep debt of gratitude to my parents, who encouraged and supported me to be all that I could be. I also had many wonderful teachers that informed and inspired me. Of many great teachers, Professor Ron Howard of Stanford University made a particularly profound impression and set me on a life long career in decision consulting. I learned most of what I know about portfolio management from colleagues at, and clients of, SRI and SDG, as well as all the great clients I had during 7 years at HP. I also must single out my wife, Kasee, and two kids, Chris and Jennie, who were shortchanged for years while I put career first. Of them all, I dedicate this book to my loving wife, Kasee.

Michael M. Menke

First and foremost, I give thanks to my family who supported my quest to create this book with my EPMC colleagues. To my Tony, who often played Mr. Mom and indulged my late nights. To our wonderful children, Jordan and Malana, who give meaning to my life. Many thanks to the business leaders who invested time to coach/mentor me through the business quagmire. And those who saw my potential and gave me a chance. Also, thanks to my core friends and extended family who are there when I need them. Lastly, most sincere thanks to those who said that I couldn't. You made me fight harder, which made me stronger; and now I am better for it. May this serve as a small inspiration for others to stay their course despite the noise. I dedicate this to book to those with big dreams for whom failure is not an option.

Diane D. Miller

There will never be enough space to thank everyone I'd like. As a start, I'd like to thank my wife, Anandhi, and three young sons, Gian, Myan, and Ashan, for allowing me to steal some quality time away from them to contribute to the book. To my mother, sister, and uncle 'Pancho' for providing me with the love, inspiration, support, and guard rails in life. To the many organizations, coaches, and leaders that saw the potential in me, gave me the opportunities to stretch, and supported me through the challenging times. To my professional mentors, teams, and reports for providing me with unique perspectives and insights. As it relates to this book, I want to thank my EPMC colleagues for the opportunity to lead the group for the last four years and for their incredible collaboration, foresight, and support. Finally, I'd like to dedicate this book to my Aunt, Monique Panchalingam, for her love, guidance, and support.

San Retna

In our lives we are all influenced, shaped, and molded by many people and experiences to become the people we are today. I am no different. The list could fill pages, but certain folks stand out for me whom I would like to thank here. First and foremost, I want to thank my Lord and Savior Jesus Christ, without whom I would be nothing. My family has stood by me, loved, supported, and encouraged me in all I do—I couldn't ask for a better clan—thank you Candice, Shaun, Marissa and Darby—my daughter, son, daughter-in-law, and son-in-law, respectively. Thank you to my grand-children who bring such joy and love into my life: Aria, Lilje, Josiah,

Benjamin, and Annabel. Thank you to my parents, David and Wanda Stratton, who gave me life and encouraged me. I dedicate this book to my soul-mate, best friend, and the love of my life—my wife, Cheryl. Honey, I love you and thank you for marrying me.

Michael J. Stratton

Life is a sum of all the choices we make. Our choices define who and what we are today. I am fortunate to have made the choice to associate with the individuals in the EPMC group. They are the finest, hard working, collaborative professionals I have known. My heartfelt thanks to EPMC for having me share and learn with you on this project. My thanks to all my colleagues at work whose support and insight helps me enhance my knowledge. My contributions to this book couldn't have been possible without the sacrifices made by my wife, Anuradha, and daughters, Sravanti and Sruti, who had to give up some of their pleasures so that I can have mine. I dedicate this book to my family, friends, and colleagues who help define who I am today.

Sarma Tekumalla

First and foremost, I would like to thank my immediate family—my wife, Kristin, and my children, Alexa, Michael and Matthew. You are my foundation and inspiration for all that I do and still hope to be. To my parents, who remain my role models to this very day, for their un-conditional and enduring love, support, and sacrifice. To my extended family, friends, and professional colleagues for the many good times we have shared and the tough times that drew us closer together and wiser as a result. To my mentors, professional peers, and associates for their advice, counsel, and insights that help me more than you know. I count my blessings each day for you all being a part of my life. I love you all very much! I dedicate this book to my dear friend Zackarie Lemelle. You have been a tremendous influence on me in so many ways and I am so very fortunate for all that you have done for me.

Mark A. Wybraniec

PROJECT PORTFOLIO MANAGEMENT

Part I

INTRODUCTION

Chapter 1

What Is Project Portfolio Management?

INTRODUCTION

"I don't understand, why aren't these projects delivering as they promised?"

This familiar cry has been heard from business leaders and project managers for some time now. Thousands of books and articles offer answers to this question, but the frustration continues. An idea that is gaining ever more traction in answering this question is Project Portfolio Management—the concept of focusing on the selection and management of a set of projects to meet specific business objectives. But when business leaders and project managers review this concept of PPM, their response is often: "This portfolio management stuff sounds way too simple. It just can't be the answer!"

However, this response itself begs a question. If PPM is so simple and self-evident, why does it have such limited traction in organizations that are apparently so in need of its help? The logic of simply reviewing all projects underway in an organization, making sure they meet business needs, align with strategy, and provide real value does *seem* self-evident. Practice and observation tells us that PPM does work, when properly implemented. Unfortunately, what our experience tells us is that a lot of the time, it's the implementation of PPM that leaves much to be desired and results in responses such as:

- "This process is too complex."
- "We don't have time to go through all this business case stuff—we need to get to work!"

- "This process is really needed for our organization's business projects, but mine are different and don't need to go through all those steps."

Apparently PPM isn't so self-evident after all. So what do we do?

Business leaders want the business to be successful. They want sound business processes they can depend upon. Project managers want their projects to be successful, so the company will be successful. So it sounds like we're all on the same page, right? Wrong. Here's where the age-old dilemma rears its ugly head for the business leader and project manager alike—there are limited resources, lots of ideas and projects, only so much time in a day and . . . oh yes, things keep changing.

This is when it becomes important for us to be able to make tough decisions: which projects do we invest in (and over what timeframe) to be successful? This requires good facts to make the right decisions. We need to be able to examine the facts when changes and issues arise that require a decision be made and acted upon. And these facts need to be weighed against our gut feel for the situation (sometimes called "experience")—by both business leaders and project managers—and then a decision made. This, too, may seem to be self-evident, but is it really? So, how do we get the facts and data we need? And how do we know we're making the right decisions?

This is where the power of PPM comes into the picture. PPM forces us to think strategically: what we want our organizations to be, and what we should be doing to get there. But it's not an easy fix. When implemented properly, PPM often requires organizational change across the business, and that can be very difficult to carry through. However, as this book demonstrates, the potential benefits for the business can be immense.

SUCCESSFUL PPM

PPM invariably changes the culture of the business because it demands we ask the hard questions. Five such questions rise to the top of the list and will be explored in depth in the chapters that follow (see Figure 1.1). Your ability to answer these questions accurately will determine how well you've implemented PPM in your organization:

1. Are we investing in the right things?
2. Are we optimizing our capacity?

Successful PPM

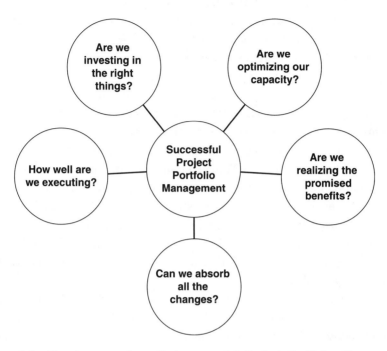

Figure 1.1   Five key questions that successful Project Portfolio Management addresses.

3. How well are we executing?
4. Can we absorb all the changes?
5. Are we realizing the promised benefits?

THE FIVE QUESTIONS IN BRIEF

Let's take a brief look at the five questions we will explore in depth later.

"Are we investing in the right things?"

Any task, activity, project, or program requires either money, equipment, material, people's time, or some combination of these. And when you look at it, the equipment, material, and even people's time can be readily converted to a common unit of measure: money. Therefore, since PPM is looking at these things as a whole, and they all take money in some form, then it only makes sense to view them as "investments." If our projects are investments, then doesn't it make sense to ask whether we're actually spending our money and time on the right things? And, so, we have the first question: *"Are we investing in the right things?"*

A sound PPM capability requires, at a minimum, four things: informed managers, involved participants (including the right level of executive sponsorship), good facilitation, and appropriate processes, systems, and tools. (Okay, that may technically be six things—we just view processes, systems, and tools as a single, integrated item—but you get the picture).

Since money is very much a limited resource, we must figure out a way to invest in the right things. This is a balancing act between the desire to fulfill the business strategies, the limited money we have to invest, and knowing when is the right time to start a project. Along with deciding which new projects deserve investment, we need to monitor the progress of active projects so that, if they're not reaping the expected benefits, they can be closed down, and their allocated capital can be recovered to apply to more beneficial projects.

However, this is not all. Businesses operate in a dynamic environment that shifts strategic objectives over time. Projects that are strategically aligned today may not be tomorrow. So PPM must also be a dynamic process. Ideally, the portfolio would be optimized in real-time (or near real-time). Also, since not all good projects can be approved immediately, what is "right" for the portfolio may not be optimal for all the potential projects competing for funding.

Foundational Tool

The Business Case

Along the way to successfully implementing PPM, we discovered that there is a foundational, and essential, tool that is often overlooked. This tool is the Business Case. It provides the necessary facts and data for understanding the value, cost, and benefit of implementing a project. It also lists the assumptions used to reach the touted conclusions, the various options considered, and the required cash flow for implementing the project.

Ultimately, the business case elicits a decision about the project, and you're given one of three choices:

- Go
- No-Go
- Wait.

One of the keys to making the best decision is understanding the criteria used to judge and prioritize projects. The company already has projects under way, and usually has a list of possible projects to add to that inventory. So how do you decide which ones to add, and when to add them? The business case is your fundamental tool for providing facts and data about each decision criterion to enable apples-to-apples comparisons to be made among projects in determining which ones should become part of the portfolio.

Lesson Learned

Even "mandatory" projects have options.

Let us share one invaluable lesson we have learned the hard way: even "mandatory" projects have options ("mandatory" projects are required to be done, maybe by law, or maybe by your CEO). Often, people will say, "We don't need to do a business case, we have to do this project because . . . " The truth we have unearthed is that there are multiple ways to meet the mandatory requirements. For example, if the requirement was to provide an efficient mode of transport, then we could meet it with a motorcycle or a sport utility vehicle (SUV). But what are the tradeoffs between these two options? Even though we may "have to do it," planning and analysis are still needed; these are accomplished effectively by producing a business case. In addition, a business case coupled with project plans enables scenario and option analysis to aid in the decision-making process.

One of the best definitions we've found for a business case is:

"A business case is a decision support and planning tool that projects the likely financial results and other business consequences of an action." (Schmidt, 2002)

In particular, note the last part of the definition. A true business case looks at more than just "the numbers." It includes financial, strategic, commercial, industrial, or professional outcomes of the project under consideration. Ideally, the business case should have more than one option from which to select, including the "do nothing" or "business as usual" option. The decision about the project needs to be made by those people with responsibility, accountability, and authority for the resources (e.g., people, tools, machines, computers, facilities) to be allocated to achieve the desired outcome.

For now, that's enough about the first of the five key questions. There will be more later, don't worry! If you can't wait to hear more about business cases and making good project investment decisions, feel free to go straight to Chapters 3, 8 and 10. So, on to the second question.

"Are we optimizing our capacity?"

This question puts into fancy words a simple concept: since we only have so much money, time, equipment, material, and skilled people, are we using them in the best way we can to get the "biggest bang for the buck?"

- Capacity optimization can also be called portfolio resource optimization. There are two key principles to understand here:
  - Optimizing resources is about *balancing* the demand for resources with the supply.
  - The primary aim of resource optimization is to create an open dialogue, based on factual analysis, between the portfolio management office and the business project sponsors (the decision makers).

Lesson Learned

Engaging business leaders in an open, fact-based dialogue is a key outcome of PPM.

Resource optimization is achieved through the balanced management of our resources. It is about understanding, managing, and balancing the demand side and the supply side of the resource management equation.

Foundational Principle

Demand-side resource management must balance with supply-side resource management.

Demand-side resource management, which concerns all the things we *need* in order to accomplish the projects in the portfolio, entails resisting the desire to control the detail. In Chapter 4 we will discuss the role of "boulders," "rocks," "pebbles," and "sand" in properly managing our resources. To ease the planning for the management of portfolio resources we group them into three categories:

- Skills (availability of sufficient people with the right skills and experience)
- Technology environment (the capacity of the computer systems or platforms to cope with the demands of the portfolio)
- Facilities (physical infrastructure, networks, office space, real estate, and the like needed to deliver projects and that will be impacted by project outputs)

Also in Chapter 4, we will seek to understand three key planning disciplines:

- Planning for skills
- Planning for the technology environment
- Planning for facilities

In effectively implementing PPM we realize we can engage four levers that help us to manage resource capacity constraints:

- Changing timescales: shifting projects within the portfolio to flatten resource demands
- Decoupling development from roll-out: helping to flatten technical resource demand
- Descoping: helping reduce the absolute need for resources
- Removing projects from the portfolio: if none of the above options are sufficient in managing resource capacity, then projects may have to be cancelled.

In supply-side resource management, which concerns all the things we *currently have* in order to accomplish the projects in the portfolio, it is key to differentiate between the organization's core competencies (those that give a competitive edge) and those competencies that can be commoditized (general skill-sets not necessarily unique to the organization). For supply constraints, core competencies are increased by training and/or recruiting qualified people from the marketplace. Commodity skill-sets are increased internally through cross-training and externally by developing and maintaining relationships with partners having different competencies and geographic footprints.

There are several ways to deal with supply-side management of the technology environment: by using an Application Service Provider (ASP) model, virtualization, or duplicate environments to better manage constraints. In handling constraints in the supply-side management of facilities we have found it beneficial to consider creative solutions such as using temporary accommodations, hotels, regional offices, or taking over a new floor in the office building.

Lesson Learned

When seeking to implement resource management for the first time, focus on a staged approach, using quick wins to build momentum and buy-in.

So, to put the question another way: "Are we getting what we are after, by using what we have, in the best way we can?" We will explore this question more in Chapter 4.

"How well are we executing?"

Doing the work of business enables us to reap the rewards. So it only makes sense that once we set plans in motion, we should check to see how well we are performing against those plans. However, as many of us have discovered through the "school of hard knocks," the world does not hold still for our plans to be executed the way we envisioned.

PPM enables us not only to know how well we are doing on our projects, but also gives us the information we need to decide what we can do to stay in tune with the demands of the marketplace and emergent situations in the business. This may involve moving people from one project to another to meet emergent demands and knowing just what the impact will be on all of our projects as well as our entire business. It also enables us to know when to stop throwing our money at projects that just aren't producing the expected results.

The world is dynamic. PPM is as well. And just as it's important to know how well projects are performing according to plan, it is also necessary to know how well PPM is performing—how mature, efficient, and effective PPM practices are in our organizations. To understand our PPM performance, we need to assess where PPM is now in our organization and what pieces are missing. Equally important is creating a clear view of this current state and gap assessment to ensure that we can progress on a defined path in adding those missing pieces. Ideally, the assessment results will show that our organization is on a process improvement path with ever increasing effectiveness toward the governance of our portfolio.

Foundational Tool

The Project Governance Process Map

One approach to establishing the clear path is the development of a Project Governance Process Map (see Chapter 5 for an example). Simple process workflow tools, such as Visio®, are underappreciated tools in the arsenal of the PPM professional. (We've heard practitioners claim to be able to "conquer the world" of PPM practice with just Excel® and Visio®.) A Project Governance Process Map is a diagram that depicts all the funding and governance steps and checkpoints that our organization has currently established for the project funding lifecycle. The completion of the map enables us to understand where we can improve the process. We discovered, for instance, that the business case for a specific project can be compelling, but unless we can see the pipeline of projects, it is difficult to make the best decision for the company in the long run.

Lesson Learned

Individual project business cases can be compelling, but we need to look at the entire portfolio pipeline to determine the best course of action for the company.

We'll look at the Project Governance Process Map in more detail in Chapter 5.

"Can we absorb all the changes?"

Ideas for new changes to our business processes, products, organizations, computing systems, and so on simply seem to have no end. However, not every idea is a good one. And not every good idea should be implemented right now.

This is what the fourth question in PPM addresses. Given the limitations of what resources we have, as we talked about in question two, and the need to track performance against plans, discussed in our considerations of question three, PPM allows us to determine what the right thing is to do at just the right time for the biggest benefit.

It gets back to having the facts in order to make good decisions.

What we have seen all too often is that we decide to move forward with a project solely on the merits of the individual project, while hoping the business can do the job. Without a way of looking at the landscape of projects, it is virtually impossible to know if a new project can even be done given the availability of our current resources for it to gain the company any benefit at all.

Another way to look at it is from a nautical standpoint. As an admiral of a fleet of ships, I make the decisions on when to launch my ships and where to send them. Oh yeah, one little detail: we haven't yet invested in a tracking system for the fleets—but we're considering it! So I have no way of tracking where all the ships are at any given time. Now back to my plans: I can look at my ships and crew for launch whenever they are ready, and then give the order to launch. Or I can look at the whole of my fleet, review my strategy and purpose for the fleet, and then deploy the right ships to the right places to effectively execute the desired job. Oh, that's right—I don't have a way to know where my ships are; . . . oops, we're definitely going to have to give that tracking system another look!

Through experience, we have found there are several different types of change we need to consider when looking at whole portfolio as well as individual projects. There's change that impacts technology, there's change that impacts physical assets (such as real estate), and then there's change that impacts people. It's this last category that really matters, as it's only people who get unsettled by change. Technology and physical assets don't have emotions. So our focus needs to be on the types of change we're driving onto people, and over what timeframe. There's clearly a world of difference between people undergoing change once a year compared to once a month.

Lesson Learned

Change may impact technology, physical assets, or people. People are the ones who get unsettled by change.

Chapter 6 outlines a fact-based methodology that enables us to look at change in terms of *what* (the degree of disruption), *when* (the timing of these changes), and *who* (both individuals and groups of people) is impacted. This methodology allows us to present our change analysis, along with recommendations, to our decision makers. Once we've started the change process and controlled the impact of change across our business, we will be able, slowly and controllably, to increase the capacity of the entire organization to handle more change.

We'll explore this question further in chapter 6.

"Are we realizing the promised benefits?"

Now that we've launched our projects, the payoff to all our hard work will just happen! What? You say that isn't necessarily so? Why not? Didn't we know what the benefit of doing the project would be? Didn't we have a way to keep tabs on the project's impact on the object of change? Didn't the money just roll in?

This brings us the final key question that PPM addresses. Or as the lady said in the old television commercial: *"Where's the beef?!"*

Effective PPM enables us to know what benefits to expect from a project and to track the realization of those benefits as the project progresses. Realizing benefits in practice is dependent on deliberate management action: staffs need to be trained to use the system and to exploit its capabilities; business processes need to be reengineered; and resources need to be redeployed. Unless this happens, the full potential benefits of our investments may not be realized. It is this problem that benefits realization management seeks to address. Chapter 7 will explore this topic further, including examining the "benefits puzzle" and "The 10 principles of effective benefits realization management."

The 10 principles of effective benefits realization management

- Benefits must be placed at the center of the portfolio management and investment appraisal processes: funding should be linked to benefits forecasts, and key stakeholders should be clear about what benefits they are buying.
- Benefits realization starts with the Business Case: ensure that the business case includes all activities and costs required to realize the forecast benefits.
- Funding allocations should be incremental, and continued funding should be directly linked to the latest benefits forecast: regular checkpoints (stage gates and portfolio level reviews) should be built in so that if benefits fall away, budgets can be adjusted accordingly.
- Where possible, "book" the benefits early: by cutting budgets, limiting headcount, and targeting unit costs, and by including them in divisional and individual performance targets.
- Optimism bias is a reality: benefits tend to be *over*stated and are often little more than unsubstantiated assumptions. Such claims must be robustly scrutinized and challenged.
- Benefits should be validated wherever possible to ensure they are realizable, by making sure that the recipients and those who will be responsible for delivering the business changes on which benefits realization is dependent agree that they are truly benefits.
- Capture all forms of value added: efficiency (both time and financial savings), effectiveness (improved performance), foundation/potential opportunity value, and the value represented by the avoidance of "things gone wrong."
- Benefits need to be *actively* managed, to ensure that forecast benefits are realized (especially important where those benefits are

dependent on business change) and to capture benefits that were not anticipated at the Business Case stage.

- Plan and manage benefits realization from a business rather than a project perspective: benefits are usually dependent on business change and may not be realized until after project deployment has been completed and the project team has disbanded.
- Utilize summary documentation and leverage the Pareto principle: short summary documents (business cases, benefits reports, and so on) convey the salient facts far more effectively than long documents.

PROJECT PORTFOLIO MANAGEMENT DEFINED

Okay, enough with the questions. So just what *is* project portfolio management? Rather than reinvent the wheel, we'll draw on what exists in the literature today for a succinct definition (Project Management Institute, 2006):

> The centralized management one or more portfolios, which includes identifying, prioritizing, authorizing, managing, and controlling projects, programs, and other related work to achieve specific strategic business objectives.

PPM accomplishes its purpose by adhering to some fundamental actions. PPM:

- Ensures that projects and programs align with the strategies, goals, and objectives of the business
- Communicates project and program details, including costs and benefits
- Manages projects and programs as a whole, providing a holistic, systems approach to business projects

---

Foundational Principle

PPM ensures the alignment of projects with strategies, communicates project details, and manages projects holistically.

Now wait a minute, you say. You've heard about portfolios, but have you heard about the different types of portfolios that can be found in business? And how are they different? Quite simply, from a management perspective, they aren't different. The bottom line is that it's all about effectively managing the work a business is doing that costs money with an eye toward fulfilling the strategic goals and bringing financial and nonfinancial benefits to the company.

The following are some of the many variations on the theme of portfolio management found throughout organizations today:

- Project Portfolio Management (naturally!)
- Application Portfolio Management
- Product Portfolio Management
- IT Portfolio Management
- Asset Portfolio Management
- Enterprise Portfolio Management
- Investment Portfolio Management
- Investment Management
- Resource Portfolio Management
- Options Management
- Pipeline Management
- Software Portfolio Management
- Governance Process

PPM is about action, so that's what we'll focus on. This book will not delve into how business strategies are developed. There are plenty of books and articles to help you do that. PPM acknowledges that strategy development is not just a linear process, and that strategy makers need feedback on how the strategy is working. This is one of the critical roles of PPM. By informing strategy makers, PPM makes strategy development and maintenance a more interactive process.

Foundational Principle

PPM is about action.

In performing its role in capacity management, PPM provides information on resource allocation and its impact and affect on strategy and the other projects in the portfolio.

As mentioned before, implementation of PPM is not easy. We hope this book will help the reader to identify blind spots when attempting to implement PPM through the authors' sharing the lessons we have learned through the school of hard knocks. This book will not go into the detail of planning a project, but should provide some critical success factors (our "aha" moments) to effectively implement PPM.

THE PPM PLAYERS AND ROADMAP

Now, you may be saying to yourself, "Yeah, I can see how PPM would work, but I can't get the whole company to agree to use it." Well, that's why we wrote this book—to help you build your story about why PPM works and, in particular, how it has worked in our organizations. We can say, without qualification, that PPM can work at an organizational, business unit, or enterprise level. Ideally, we know it works best if it can be implemented enterprise-wide, but we have not seen this happen very often in real life.

In fact, this brings up a topic we will cover more in Chapters 8 and 9— there is, and is not, just one portfolio. "*What?!*" You heard right. From the perspective of the enterprise, all projects are in the one enterprise portfolio. However, each business unit and organization has a piece of that portfolio that they manage using the PPM process, and each of those pieces is a portfolio as well (the business unit and/or organizational portfolio). Most companies do not attempt to run all their projects at the enterprise level; that would be crazy. It turns out that PPM is actually a set of tiered portfolios (as opposed to what some might say are really "teared" portfolios, given the work involved). What determines the movement of projects from one portfolio to another is thresholds (see Figure 1.2).

Foundational Principle

There is and is not just one portfolio. It's a tiered thing.

The critical factor to understand here is that you don't need to implement PPM at the enterprise level to see the benefits of PPM. Sure, it may be easier if everything is aligned throughout the enterprise, but running a company would be easy if things never went wrong. We'll get into the details of how this all works a bit later. If you want to see right now, just turn to Chapters 8 and 9 and it will make more sense.

Project Tiers

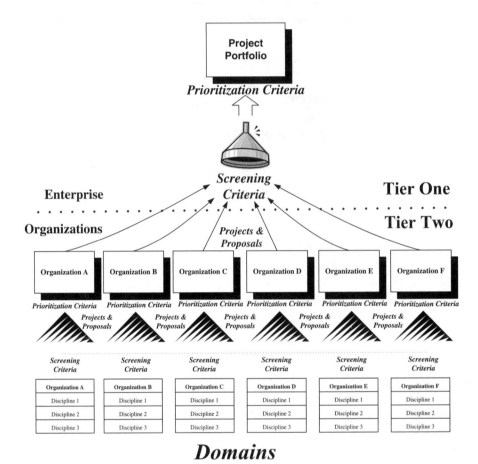

Domains

Figure 1.2   Projects are screened and selected in tiers, first at the business unit or organizational level, and then at the enterprise level (assuming the organization practices enterprise PPM).

Foundational Principle

You don't need to implement PPM at the enterprise level to see the benefits of PPM.

THE PPM PROCESS VIEWS

We may view the PPM Process from several different angles. We earlier looked at some of views of portfolios with relation to the enterprise. Now let's take a look at the process itself. Figure 1.3 shows the PPM Process in a general flow without specifying roles and responsibilities. As you can see, the process is iterative until we get to the closeout of the projects. Figure 1.4 shows the PPM Process in a "swimlane" format detailing specific products and deliverables, as well as responsibilities. We will go over the swimlane chart in more detail in chapters 8 and 9.

A FEW MORE QUESTIONS TO GET THE MENTAL SYNAPSES FIRING

Whenever you read a newspaper or magazine article you will see the author attempt to answer six key questions, also known as the "Five Ws and One H."

- Who?
- What?
- When?
- Where"
- Why?
- How?

It only seems appropriate in exploring what PPM is that we look briefly at the "Five Ws and One H." Here goes.

Who?

Who can really use the PPM process? The answer:

- The "C-Level" executives
  - CEOs: Chief Executive Officers
  - CFOs: Chief Financial Officers
  - CIOs: Chief Information Officers
  - CTOs: Chief Technology Officers
  - CSOs: Chief Strategy Officers
  - CPOs: Chief Portfolio Officers

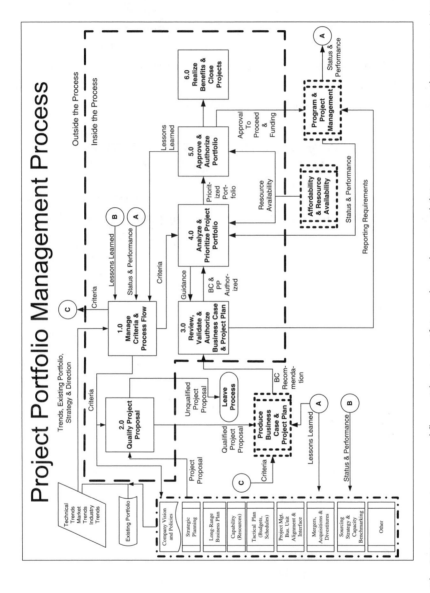

Project Portfolio Management Process

Technical Trends / Market Trends / Industry Trends

Trends, Existing Portfolio, Strategy & Direction

Outside the Process

Inside the Process

Existing Portfolio

1.0 Manage Criteria & Process Flow

2.0 Qualify Project Proposal

3.0 Review, Validate & Authorize Business Case & Project Plan

4.0 Analyze & Prioritize Project Portfolio

5.0 Approve & Authorize Portfolio

6.0 Realize Benefits & Close Projects

Program & Project Management

Affordability & Resource Availability

Produce Business Case & Project Plan

Leave Process

Criteria

Lessons Learned

Status & Performance

Project Proposal

Unqualified Project Proposal

Qualified Project Proposal

BC Recommenda-tion

Guidance

BC & PP Author-ized

Criteria

Prioritized Port-folio

Lessons Learned

Resource Availability

Approval To Proceed & Funding

Status & Performance

Reporting Requirements

Status & Performance

Lessons Learned

Criteria

B

A

C

A

C

A

B

Company Vision and Policies

Strategic Planning

Long-Range Business Plan

Capability (Resources)

Tactical Plan (Budgets, Schedules)

Project Mgt. Bus. Unit Alignment & Interface

Mergers, Acquisitions & Divestitures

Sourcing Strategy & Capacity Benchmarking

Other

Figure 1.3   The PPM Process in a general flow without specifying roles and responsibilities. As you can see, the process is iterative until we get to the closeout of the projects.

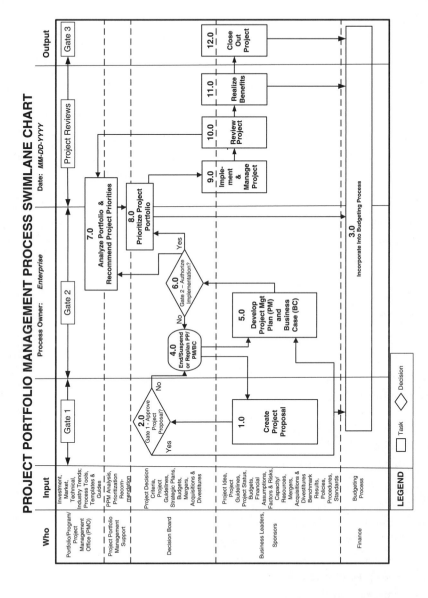

Figure 1.4 The PPM Process in a "swimlane" format detailing specific products and deliverables, as well as responsibilities.

- Non-C-Level executives
- Department heads
- Managers
- Supervisors
- Portfolio managers
- Senior project managers
- Project managers
- Program managers
- Lead engineers
- Systems engineers

What?

What should you use PPM for? The answer, managing:

- Multiple projects
- Multiple programs
- Assets
- Software applications
- Investments
- Resource allocation
- Capacity
- Products

When?

When should PPM be used? The answer:

- You have more than one project or program
- A decision needs to be made about:
  - Ideas or proposals moving to the business case and detail planning phase to compete for a slot in the portfolio as a project or a program
  - Projects or programs going forward
  - Projects or programs being "killed" or put "on hold"
  - Resource allocations are at issue between projects or programs
  - Strategies change
  - Business conditions change
  - The market changes
  - Mergers
  - Acquisitions

- Divestitures
- Joint ventures
- Buying, building, decommissioning, or disposing of facilities, equipment, or material

Where?

Where is PPM used? The answer:

- Nonprofit businesses
- For-profit businesses
- Government agencies and departments
- Universities and colleges
- Utility companies
- Investment firms (naturally!)
- Law firms
- At the enterprise level
- At the business units level
- At the organizations level
- At the discipline level

Why?

Why use PPM? The answer:

- PPM accomplishes its purpose by adhering to some fundamental actions. PPM:
  - Ensures projects and programs align with strategies, goals, and objectives of the business.
  - Communicates project and program details, including financial costs and benefits.
  - Manages projects and programs as a whole. It's a holistic, systems approach to business projects.

How?

How do you decide whether to use PPM or not?

- Engage in conversation and discussion with:
  - Executives
  - Managers

- Project managers
- Program managers
- Subject matter experts
- Develop a business case with options for managing the items under "When?"

CHAPTER SUMMARY

Foundational Principles

- Demand-side resource management must balance with supply-side resource management.
- PPM ensures alignment of projects with strategies.
- PPM communicates project details.
- PPM manages projects holistically.
- PPM is about action.
- There is and is not just one portfolio. It's a tiered thing.
- You don't need to implement PPM at the enterprise level to see the benefits of PPM.

Foundational Tools

- The business case
- The project governance process map

Lessons Learned

- Even "mandatory" projects have options.
- Engaging business leaders in an open, fact-based dialogue is a key outcome of PPM.
- When seeking to implement resource management for the first time, focus on a staged approach, using quick wins to build momentum and buy-in.
- Individual project business cases can be compelling, but we need to look at the entire portfolio pipeline to determine the best course of action for the company.
- Change may impact technology, physical assets, or people. People are the ones who get unsettled by change.

Part II

PROJECT PORTFOLIO MANAGEMENT: A STORY

Chapter 2

Introduction

This story will be used as the backdrop to provide recommendations to address some of the ills described. Although we do not profess to offer one-size-fits-all suggestions, we all just might find the pearl of wisdom we need to make a difference.

The Cast

ACME RETAIL, INC.

Located in the Midwest, Acme is a $6 billion (bn) consumer retail company. The company has a portfolio management office, headed by John Durham. At nine months of John's 'tenure they have ninety-seven projects, nineteen portfolio programs, and eighteen divisional programs being overseen by the PMO.

JOHN DURHAM

John is the main character in the story. He is the vice president of portfolio management for Acme Retail, Inc. He is relatively new in the position. At the start of the story he has been at his new job for six months. He attended college in the New England area with Bill Smith, where they were roommates in a fraternity. John and Bill graduated ten years earlier.

BILL SMITH

Bill comes to the aid of John as he struggles with PPM. He is a PPM practitioner and head of the Enterprise Portfolio Management Council (EPMC). Bill is our story's expert. He has gathered a lot of experience while implementing and managing PPM since graduation—it's his passion.

CHRIS (CHRISTINE) JONES

Chris is Bill's colleague at the EPMC. Chris is also a PPM practitioner at an information technology company and has lots of relevant experience that helps John.

LEO PALMER

Leo is the Chief Operations Officer at Acme Retail. He has come to John wanting to know what was happening and what they should do to turn things around. He has asked John for "a nice little chart" he can take to Hannah the CEO, that will clearly show that we need to slow things down a little. He needs more than a general assertion that our people are reeling from all these initiatives. Without hard data, he expects Hannah to bite his head off and hand it back to him on a platter. She views John as their portfolio management guru. She is generally concerned about change and wants to know how much change is too much.

HANNAH CHOI

Hannah is the CEO of Acme Retail. The daughter of Korean immigrants, she has risen to the top after having started her career stacking shelves in one of Acme's stores. She is very results-orientated and has great ambitions both for Acme and herself. If she can see how PPM can benefit the business, then she will adopt it with vigor.

JORGE GARCIA

Jorge is John's office assistant. He is very efficient and effective but a little on the formal side in his dealings with his boss.

ALEX

A member of the EPMC, Alex works in a major pharmaceutical company.

THE TIMELINE

Figure 2.1

Story Timeline For Book

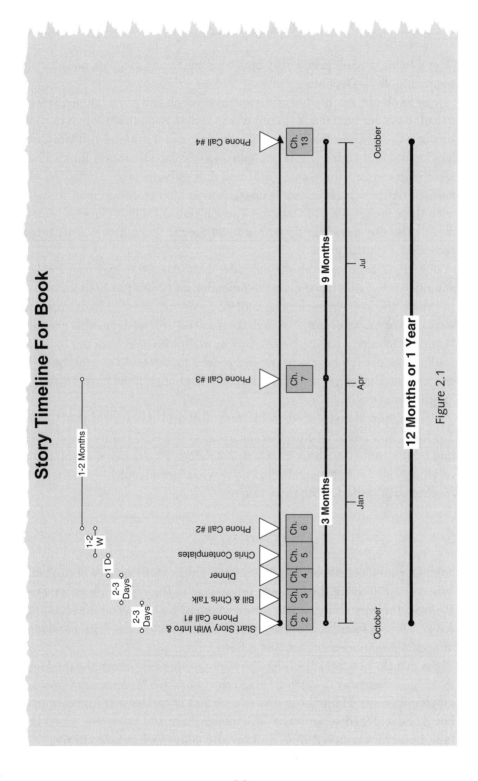

Figure 2.1

29

10 YEARS AGO . . .

"Hey, John, catch!" yelled Bill Smith as he threw a bullet pass to his roommate John Durham.

The football flew in a beautiful spiral through the crisp fall air. It was one of those perfect New England days tourists hope to see when coming through the area on fall foliage expeditions. But John and Bill were not looking at the weather, or at the splashes of color all around them. They were engaged in "The Game." Bill and John's fraternity, Alpha Chi Mu Epsilon (ACME, nicknamed "Tune-Townies"), were going head to head with their archrivals Chi Omicron Tau Epsilon (XOTE, nicknamed "Coyotes"). The Coyotees had 28 points to 24 for the Tune-Townies with ten seconds left in the game.

John had three steps on the defender as he turned to look over his left shoulder. The ball zipped over the defender and John put both hands up to cradle the ball. Whack! The ball bounced off John's hands, up and forward. John dove and snagged the football inches from the ground. "Umph!" John cried as he hit the turf hard, but he held onto the ball. He couldn't believe it—he was in the end zone! *Touchdown!* The gun sounded the end of the game and the "Epsilon Alphabet Bowl" was over for another year.

By the time John leaped to his feet, Bill and the rest of the "Tune-Townies" had engulfed him in raucous laughter and manly hugs. They had finally won after three long years of defeat. The trophy—if you could call the welded car parts and who knows what else a trophy—was theirs to display with pride for the next year.

PRESENT DAY . . .

John Durham looked out his office window at the still surface of the lake, which was reflecting the brilliant fall foliage in the bright afternoon sun. He loved the view from his office from which he had watched the changing seasons for six months. It came with his new job and title—vice president of portfolio management for Acme Retail.

Six months and he still felt like he was spinning his tires in the mud and making no headway in getting this Project Portfolio Management process moving forward. He had come to the job full of excitement and promise, but as he watched some leaves drift down from the trees that lined the lakeshore, he wondered if he could really deliver what he'd offered.

The intercom buzzed, snapping him out of his brooding. "Yes," he said, pressing the button on the box.

"A Mr. Smith is on the line for you, Sir. He says he's an old college buddy. Do you wish to take the call?" his ever-efficient office assistant Jorge asked.

A smile crept onto John's face and into his voice as he replied, "Thank you, Jorge. I'll take the call in here. Oh, and I'd like you to hold my calls for about fifteen minutes."

"Absolutely, sir," Jorge responded.

"Bill, you old coot! How are you? Been passing any footballs lately?" asked John in an excited voice.

"Not lately," laughed Bill. "How about you?"

"No. And I haven't been catching any either. I was just looking out the window at the fall foliage and, with you calling me out of the blue, it takes me back to ten years ago and our victory in the Epsilon Alphabet Bowl," John said.

"I still remember that day. Boy, what a feeling . . . and what an ugly trophy!" exclaimed Bill. "Say, I was just calling to let you know I'll be in town next week and was wondering if we could get together for dinner."

"That would be great, Bill. I would love to see you again."

"I hear you were recently advanced into the executive ranks at Acme. How's it going?" asked Bill, genuinely interested.

"Yep! Six months on the job today," said John a bit more enthusiastically than he truly felt. "What are you up to these days?" he asked, sidestepping the question.

"I moved into a new position as vice president of a project management office, running the company's project portfolio a couple of years ago. I also helped found a PPM group called the Enterprise Portfolio Management Council—maybe you've heard of it? The EPMC . . . ."

"Well, now you mention it, I think I've read about the EPMC in some of the stream of marketing material that floods across my desk. That sounds great—and what a coincidence that we both ended up in the same area after all these years," exclaimed John. "Since you're in town next week, by all means let's get together and catch up."

"That's a deal," said Bill.

"So, what kind of projects is your company involved with?" asked John.

"We try to run a balanced project portfolio. So, we have a variety of projects. It been a bit of a roller-coaster ride but I think we have a pretty well-oiled system in place now," stated Bill matter-of-factly.

"Really," said John. "I wish I could convince my colleagues to run a balanced portfolio! If the truth be known, it's been a tougher challenge than I expected getting PPM running well. What would you say was the critical thing to get right, Bill?"

"Well, it wasn't an overnight, big-bang implementation," said Bill. "It took a lot of time, thought, and planning to get it rolling. I had a tough start but what I found worked well in persuading the management team that PPM was worth doing was writing out the answers to five questions. It really helped me to formulate a clear roadmap for implementing PPM."

"So what are the questions?" John asked excitedly.

"The first one is: Are we investing in the right things? The second question is: Are we optimizing our capacity? Third: How well are we executing? Fourth: Can we absorb all the changes? And finally: Are we realizing the benefits? I got the ball rolling by asking myself why we started PPM in the first place."

"Bill, those are great questions! I'd like to go over them in more detail with you over dinner next week."

"Absolutely," said Bill. "I'll find a restaurant near my hotel, make reservations, and give you a call on the particulars."

"Sounds great," replied John. "I look forward to catching up with you— and going over those questions! See you next week."

"You bet! See you then," Bill said as he hung up the phone. "I better get Chris to look into the business situation with Acme before the meeting," thought Bill. He had a funny feeling the dinner would not be so much about catching up, as it would be about going over answers to those questions.

LATER THAT AFTERNOON . . .

John was thinking that Bill's call might have been just what he needed. He pulled a piece of paper out of the printer tray and started to write down answers to the five questions. He wanted to get his thoughts together before meeting with Bill. It sounded like Bill had lots of useful experience of implementing PPM and he wanted to clarify what was really causing progress to stumble.

John started writing. Before long, his hand was flying over the pages. He sat in silence and scribbled away. These are his notes.

WHY DID WE START PROJECT PORTFOLIO MANAGEMENT?

- We wanted to handle the flood of project requests coming in from all areas of our business.
- Without any way to measure the value of projects to our organization, we were queuing everything up on a first-come, first-serve basis; not exactly the best way to set priorities.
- We also had a reputation for taking care of whoever yelled the loudest—and we had a lot of screamers!

ARE WE INVESTING IN THE RIGHT THINGS?

- I can't say that we are with *confidence*.
- We formed a Decision Review Board (DRB) charged with setting the direction for the project portfolio. This seems to have helped us get moving in the right direction and it seems to be effective.
- The DRB was charged with the responsibility for determining which projects were worth taking on and determining whether active projects were worth continuing.
- We have the commitment from the executive leadership team— they committed the top people in the organization and everyone on the DRB seems to take their role seriously.
- We've prioritized work and actually stopped a few projects that we did not feel were strategically aligned.
- Our projects are usually approved on financial merits. And once we secure funding, then we get the green light to start working.

ARE WE OPTIMIZING OUR CAPACITY?

- Not sure of the answer to this!
- We have a problem of overcommitting and underdelivering.
- We have a lot of smart people and many have great ideas but they all need resources to execute.
- No matter how many people we hire, we never seem to have enough to get the work done.
- While it is important to ensure we have the money to commit to a project, we also need to make sure we have the right people to staff the work and the facilities to support them. This *minor*

detail is sometimes overlooked, yet it may be one of our major reasons for project failure.

How well are we executing?

- I honestly don't know!
- I can't talk about Project Portfolio Management without addressing our ability to execute.
- Project Management (PM) is part of it, I know.
- PM is all about planning the work and working the plan. But we seem to dive right into the "doing" without much planning. Planning seems to be too hard for our people. It requires a solid understanding of the subject area, strong project management skills, and time to actually think about what needs to be done.
- We don't appear to invest much in the development of project managers. Nor do we give the project managers who have the skills the right autonomy and authority to do what they do best.
- We seem to fall short on some of the project management basics.
- Except for the very large capital expenditures, most project funding is absorbed as part of the various departmental budgets. Borrow a little here, beg a little there.
- Who knows what was spent on a particular project when the spending is scattered to the four winds? We seem to have trouble connecting our project budgeting in our financial systems. How can I fix that?
- I have been a project management professional for many years and I thought that I had seen it all in my career, but here I have been introduced to the practice of creative project management. It is amazing that project budgets (assuming that you have a budget) were always equal to the final spending and very few projects ever went beyond the committed delivery date. Seemed a little fishy to me, but the mystery was rather easy to solve. It seems there is really no formal project accounting in our financial systems, so it is easy to get creative with funding. And if you do not commit to a delivery date until a week before you are done, it is a pretty easy to hit that target date!

Can we absorb all the changes?

- I can only recount anecdotes overheard on failed attempts to introduce change, but I can't provide an assessment of the current state.
- Change is a necessary part of life. I am a strong supporter of change—but managed change, not change for change's sake. We must understand that too much change in any area can significantly impact the ability for people to get their jobs done effectively.
- Sometimes we seem to be overly zealous about making things happen. So much so that we sometimes forget about the people who are impacted by all these whiz-bang new gadgets and processes that we put in place.
- Every organization has a different level of change that it can absorb and we probably need to factor this into our plans.
- Sometimes we implement a solution only to change it a few months later. It seems the left hand does not know what the right hand is doing. This is a big problem.
- I have noticed that some of our people seem confused and frustrated and may try to hold off on adopting new systems and practices until they are tried and true. Some of our best people have also left us to pursue opportunities in environments that are a little less chaotic.

Are we realizing the promised benefits?

- Again, I have no idea!
- We are working really hard, but are we reaping any rewards from what we are doing?
- Once projects are implemented, we are off and running on the next thing. We have not historically had the staff to monitor benefits. The attitude of most in the organization is that once the project is done, it is done.
- We have not adjusted budgets based on stated reductions in cost, nor have we held anyone to commitments made in increasing revenue or profits.
- I don't have anyone on my team to help in this area, so it has been something that we have not even attempted to grapple with yet.

FINAL THOUGHTS

- How can I communicate the merits of Project Portfolio Management? Making sure that we are doing the right things at the right time should be self-evident. Or is it?
- We have had small bursts of success.
- I seem to have trouble sustaining the processes that do work.
- I am not ready to throw in the towel yet.
- I just don't know where to start.
- I need to have a proactive plan to get us back on track!

Chapter 3

Are We Investing in the Right Things?

Shortly after his call with John, Bill contacted one of his colleagues in the EPMC, Christine Jones. She had almost as much experience of PPM as he did and he had recruited her into the EPMC right after it was founded. Chris (as she preferred to be called) was a genius at research and so he asked her to look into the situation at Acme. Two days later she got back to him.

"Hi John," Chris said as she entered the office. "I've found some very interesting things about Acme and your college buddy John. You two should have a really interesting evening."

"That's great," said Bill. "So what exactly have you found out?"

"Well, did John mention that he was the VP for Project Portfolio Management at Acme?"

"Yes, he did—said he's been there for six months", replied Bill.

"Correct. He's been in that role six months," said Chris. "The press release announcing his appointment mentioned that a primary goal would be to get projects moving through their pipeline. They have aggressive growth plans that are generating a lot of projects to support the growth goals. However, I'm guessing that he may be running into problems since Acme is still rumored to be well behind on most of its big projects and initiatives."

"Well, that probably explains why he's so keen to meet with me," said Bill.

Chris continued, "All too often, the people put in charge of portfolio management don't know much about what it is and how to get it done.

Many of them are coming from operations or, as in John's case, finance and accounting. That's a very different area."

"Well, that is interesting. I had assumed that he had as much experience of portfolio management as we have. That would certainly explain why he could be running into trouble," said Bill.

Chris nodded and said, "I bet John is wrestling with the classic first question for novice portfolio managers—how do I know if we are investing in the right things?"

"Oh, I'm sure you're right. And now that you've found out he comes from a finance background, I bet that's the case. If I know Bill, he's probably ended up bogged down in the detail, convincing himself that most, if not all, project ideas make sense when looked at from a strict cost-benefit perspective. He needs to lift his focus. You know, this reminds me of what happened with me at my last company."

Bill pushed his chair back and wedged his hands behind his head. Chris recognized the sign and slumped down into one of the chairs.

"When the CEO showed interest in exploring portfolio management, I sold it to him on the basis that we would be able to ensure strategic alignment for every project. I still remember the day. I drew the triangle diagram for him and he loved it." Bill leant forward, pulled out a piece of paper, and sketched this figure:

Bill continued, "At that time, the CEO was all about strategic alignment, so we started with that as our focus. The project portfolio was

Figure 3.1

derived straight from our strategy—in fact, we initially called it 'strategy translation.' The projects were all aimed at filling gaps in skills and capabilities that we needed to implement our strategies. Then, after about a year, our next step was about portfolio balancing. We were less focused on maximizing financial return because we initially were looking at the service aspects of the organization and the return from some of their strategic initiatives was not easy to quantify. So for us, doing the right things was strategy first, balance second, and return third."

"That's a great example—you should tell John," said Chris. "I researched a very different situation for the EPMC a year or so ago. It was a basic commodity materials company and a new CEO had taken over with a stated mission to diversify and take them into higher-return, less cyclical businesses. At the same time they got a new VP of R&D who wanted to show that he could manage the R&D project portfolio well. The trouble was that declaring an aim of diversification into higher-return businesses was not specific enough to allow any meaningful strategic alignment. Virtually any project not directly in the core business that had a high return and was in any way related to the existing business could be pronounced to have 'strategic fit.' Strategic alignment became almost meaningless as a test of doing the right thing. In the end, they focused heavily on finding the projects with the highest returns, but were careful to assess the risk as well. So you could say that in determining the right R&D projects they put maximum return first, balancing risk and return second and strategic alignment was a distant third. A very different situation, but a real case all the same."

"What about all the portfolio work Alex did with that pharmaceutical company?" asked Bill.

"Oh yes, that's another great example. The pharmaceutical company expressed their strategy in terms of choosing therapeutic areas to focus on, and then looking for high return projects—blockbusters if possible—within those areas. They also had a very structured approach to high-quality business cases for each proposal that involved quantifying the return and risk for different resourcing levels according to where they are in the development process. In fact, some of them have applied very sophisticated approaches to optimize their expected returns using the so-called 'efficient frontier' technique. Pretty fancy stuff, and it sometimes pays off very big for them with returns in the billions of dollars," said Chris. "I'm not sure whether to say they apply strategy first or optimize return first, but they sure put a lot of work into optimizing their returns.

They even go so far as to take money from some less deserving projects and give it to others that have a better use for it."

"What about projects other than R&D?" asked Bill. "Do we have some good examples from the EPMC membership?"

"Well there was that service firm where one of our newest members is employed," said Chris. "Didn't they have IT projects, marketing projects, facilities projects, new service development, real estate, customer acquisition, and other stuff all in their portfolio?"

"Yes that's right," said Bill. "But they were still a fairly small organization compared to many of our EPMC members. Do we have some really big company examples?"

"Actually, there is a great example from a very large financial institution that provides credit cards and other financial services. They have developed a Corporate Portfolio Management process based on what they call 'Investment Optimization (IO).' They have been very careful to isolate the funds they need to run the business from their discretionary investment portfolio. They insist on all discretionary projects going through their IO process. This is really an enterprise-wide implementation, involving nine business units, eight functional groups, and over fifty geographical markets. In fact, it started in their international card business and spread out from there. They use it for every type of investment and claim it impacts over five thousand investment proposals annually, worth over five billion dollars."

"Wow that's huge," said Bill. "But does it really work? What's the evidence?"

"There must be something to it because they let Anand Sanwal (2007) write a book about it, but only after applying for business process patents. And their process has been benchmarked and audited by several prestigious external organizations like the CFO Executive Board. They also won the Grand Prize in the *Baseline* magazine 2005 ROI Leadership Awards with an estimated ROI on their IO process of over 2,700 percent! That estimate was produced by an independent group so it seems pretty compelling to me," said Chris. "Sharp & Keelin (March-April 1998) wrote a *Harvard Business Review* article claiming that portfolio management had gained them over two and a half billion dollars!"

"You know, why don't you join me for dinner with John? Are you free that night?"

Chris pulled out her blackberry and flicked through her calendar. "Yes, I'm available that evening. If you really think that it's worth the travel, I would be happy to come along. And why don't you review our EPMC Working Document on Portfolio Investment before you come over. I think that might help John as well."

"Absolutely—and thanks! I have a feeling that this could become another great case study for the EPMC. When I knew John back in college, he was a really decent guy and I'm sure would be very happy to help us if we help him." Bill was always looking for opportunities to showcase the EPMC.

"That would be great, Bill. We don't have a retail case study and if we could help John right at the start of his PPM journey then we could really help him make a big difference. I'll get a package together of our thinking on optimizing the balance of a project portfolio." Chris stood up and made for the door, but then she stopped and turned around to face Bill. "You know, every time we discuss this piece of the PPM puzzle it never ceases to surprise me how, at its heart, this is so simple."

EPMC WORKING DOCUMENT ON PORTFOLIO INVESTMENT

Portfolio management is a key step in the implementation of strategy. Strategy articulates a vision for the future of the organization that usually requires the development or acquisition of new organizational assets or capabilities. Developing and/or acquiring these takes investment of financial and human resources, both of which are scarce resources. Most organizations have far more good investment opportunities than resources to fund them. Portfolio management is about the prioritization, timing, and resourcing of these investment "projects."

Given limited resources and the need to select the "best" set of projects that we can afford, how should we determine what is the "right" portfolio of investments? This is a multifaceted question in a dynamic environment. The "right" portfolio should enable strategy, deliver a good return on investment, and be balanced in supporting the various objectives of the organization (e.g., strategic success, financial success, high growth, risk management, timely execution, employee satisfaction, customer satisfaction). As if this were not already complicated enough, these objectives evolve over time and what is strategically aligned today may not be tomorrow. So portfolio management must be a dynamic process, and in the ideal state be optimized in real-time (or near real-time) as a few leading practitioners are striving to achieve today. Also, since not all good

s can be approved immediately, what is "right" for the portfolio will
optimal for all the potential projects competing for funding.

The definition of "right" also depends on who is determining the answer, so portfolio governance becomes important. Most organizations establish a portfolio decision review board (DRB), also called portfolio approval committee, with responsibility and authority to decide on the approved portfolio. If the portfolio in question is absolutely central to the future of the organization (e.g., the drug development portfolio in a pharmaceutical company), then the portfolio DRB is usually very close to the executive leadership of the organization (e.g., Head of Development, Head of Marketing, CFO, Heads of Key Businesses). If the portfolio in question is important but not absolutely central to the future of the organization (e.g., IT project portfolio in most organizations), then the portfolio DRB will consist of senior executives usually a level or two down from the top executive leadership. In any case, the "right" portfolio will depend on the experience, intuition, and judgment of the members of the DRB.

Extensive benchmarking of Project Portfolio Management (PPM) has found three overarching objectives: strategic alignment, maximum return, and strategic balance. Most organizations using portfolio management are striving to achieve two and usually all three of these objectives. Strategic alignment means that the approved portfolio supports the strategy of the organization. Maximum return means that the approved portfolio achieves the best aggregate financial outcomes (or public benefits) relative to the aggregate investment required. Strategic balance means that portfolio has an appropriate mix of projects considering the multiple objectives and mandates of the organization. It should be evident that these three objectives themselves are not completely aligned and so the portfolio DRB has to make trade-offs among them.

In addition to these three major portfolio objectives, project risk and timing are important considerations in selecting the portfolio. Risk has at least three constituents: technical risk, implementation risk, and commercial or benefit risk. Technical risk is a measure of whether the project's technical objectives can be achieved. Implementation risk is a measure of whether the organization can implement and use the technical deliverables if they are achieved. This is often a function of the organization's ability to absorb change. Benefit risk is whether the ultimate desired strategic, financial, or social benefits will materialize if the other two risks are overcome. These are judgment calls that reflect our confidence in the ability of the technical or project organization to achieve their deliverables, of the operating organization to implement these deliverables to provide new products and services, and of the end users actually to use (and even pay for) these products and services to realize the ultimate benefits desired. Since not all projects can be resourced immediately, adjusting the starting dates and resource intensity (i.e., run rate) of projects is an important consideration in crafting a portfolio that is feasible given financial and human resource constraints.

Strategic Alignment and Portfolio Management

Strategic alignment has been identified as one of the three overarching objectives of portfolio management, the other two being maximum return and strategic balance. Yet many people are unclear on exactly what strategic alignment means, let alone how to achieve it. The most common statements are that strategic alignment means the projects in the portfolio "fit" or "support" the strategy, which simply passes the buck to what fit and support mean in this context. Furthermore, this requires that the strategy must be articulated with enough clarity that this alignment/fit/ support can be tested; however, in many organizations strategy is only defined at a very high level that does not lend itself to detailed testing of alignment.

One explanation of strategic alignment is that of Cooper and colleagues (1998) in their book *Portfolio Management for New Products*. There they describe strategic alignment for new product portfolios as follows:

- All active projects are aligned with the business strategy.
- All active projects contribute to achieving the goals and objectives set out in the business strategy.
- Resource allocations—across business areas, markets, and project types—truly reflect the desired strategic direction of the business.

The mission, vision, and strategy of the business must be operationalized in terms of where the business spends money and which investments it makes. We believe that these basic ideas apply not only to new product portfolios but also to project portfolios in general and to the corporate portfolio as a whole.

First, strategic alignment should mean that the projects in the portfolio are both necessary *and* sufficient for the strategy to succeed. Necessary

[handwritten margin note: Necessity + Sufficient interdependent — need both for "Alignment"]

means that the project is required for the strategy to succeed; this is essentially what Cooper and colleagues call *strategic fit*. Sufficient means that all the projects that the strategy requires to be successful have been approved and are in the active portfolio; this is similar to what Cooper and colleagues call *strategic contribution*. Prior definitions of strategic alignment tend to emphasize necessity but not sufficiency; however, if the portfolio of projects is not sufficient for the strategy to succeed, it will fail and the "alignment" of the other projects is useless. We believe that the application of both of these tests together substantially improves the definition of strategic alignment. Finally, Cooper and colleagues introduce a third test of strategic alignment for new product portfolios, *strategic priorities*. This basically asks whether the breakdown of your spending is consistent with your stated strategy.

How can strategic alignment be achieved? Cooper and colleagues identify three general approaches: Top-Down, Bottom-Up, and a blend of Top-Down/Bottom-Up. Top-Down begins with the business's vision, goals, and strategies and then "translates" this into investment initiatives and/or resource allocations. They identify two general approaches for doing this: road-mapping and strategic buckets. Road-mapping translates the business strategy into the strategic initiatives and investment programs required to execute the strategy; road maps normally also have a time dimension. Strategic buckets translate the strategy into spending categories designated for different types of investment projects. Projects are then created and prioritized within these strategic buckets. These two approaches can be used separately or together.

Bottom-Up begins with a set of investment opportunities that can arise from anywhere in the organization but then must be screened so that the best ones rise to the top. This requires that strategic criteria must be built in to the project evaluation and selection process. The Bottom-Up approach, by starting with individual projects and programs, may miss some projects that are necessary for the success of strategy and thus could lead to an insufficient portfolio. The Top-Down and Bottom-Up approaches are fundamentally different in philosophy, implementation, and the portfolio of projects that emerges.

Top-Down/Bottom-Up combines the two approaches and thus hopes to overcome the deficiencies of each individually. It begins at the top with strategy, road maps, and strategic buckets, but also proceeds from the bottom with creation, evaluation, and selection of the "best" projects. By starting with the Top-Down approach, it is more likely that the resulting portfolio will include the projects that are essential for strategic success. Then the two sets of decisions are reconciled via multiple iterations.

The most straightforward way we know of is to generate the investment portfolio from the requirements of the strategy to be successful, based on gaps in the capabilities and assets necessary to be successful. This is a form of the Top-Down approach. AAA of Northern California, Nevada, and Utah calls this "strategy translation." In this approach, alignment is built in to the portfolio because the portfolio is created from the programs and projects the strategy needs to be successful. It has often been said that strategy only becomes real when the required resources are assigned to implement the strategy; thus strategy translation is an important aspect of making the strategy real.

Strategy translation will yield two classes of "aligned" projects: those essential to the success of the strategy (without which the strategy will fail) and those that enhance the success of the strategy (but are not essential for its basic success). Projects that are essential for the success of the strategy should be given the highest priority. If resources to execute the portfolio are scarce, we can make tradeoffs among the nonessential projects in the portfolio, using balance and return as prioritization criteria. However, we should also examine the balance of the entire portfolio (essential and nonessential projects) as well as the return of the "essential" projects. If the total portfolio is poorly balanced or if the "essential" projects have poor returns, this casts doubt on the viability and desirability of the strategy and the rationale for the strategy should be revisited.

Some possible metrics for strategic alignment include percent of projects/programs, percent of portfolio spent, and percent of portfolio benefit explicitly aligned with strategy and strategic objectives.

Maximum Return

The next major objective in portfolio management that contributes to investing in the right things is maximizing the portfolio return. In a world of unlimited resources, the value of each individual project could be optimized and then all projects executed at their optimal resource level. However, given the reality of a resource-constrained world, some low-value but still desirable projects may need to be dropped or postponed, others stretched out in time, and yet others done with fewer resources so the most valuable projects can be done in an optimal fashion. Most people have a tendency to avoid hard decisions; as a result portfolios are often overcrowded with too many projects, each underresourced to achieve its optimal outcome. This is rarely the best portfolio, or even close. Many organizations have found that it is much better to focus their portfolios on fewer, better-resourced, high-payoff projects that can be executed swiftly and achieve all of their objectives. Making these trade-offs is the essence of finding the portfolio with maximum return.

Finding an analytical solution to optimize portfolio returns is enormously complex. However, some very good approximate methods have been developed and applied to portfolios in the pharmaceutical and oil and gas industries. We shall use these to illustrate the approach. The essence is to use the wisdom and experience of the project teams to develop three alternative project plans: one the current approach as planned, another the minimal viable version of the project using fewer resources (e.g., 15–30 percent fewer), and third the best conceivable version of the project using more resources (e.g., up to 25 percent more). Some projects will suffer very little benefits reduction when using minimal resources, while others may offer very large benefits gains by getting additional resources. Portfolio optimization aims to take resources from projects that will lose the least value and transfer them to projects that can gain the most. Note that this perspective can only be applied at the portfolio level, since it is intrinsically based on moving resources among a collection of projects that have different sensitivities to gaining or losing resources.

Strategic Balance

There is an old adage, "Nothing ventured, nothing gained." A modern version is that higher returns generally require taking higher risks. Therefore, one approach to achieve a high return portfolio would be to fill it with high-risk, high-return projects. Yet this is not the way most organizations want to or should behave. A maximum return portfolio may well be a very risky portfolio. Conversely a minimum risk portfolio may suffer from meager returns. Intelligent managers seek to strike a balance of risk and return. And this is only one of many such trade-offs a portfolio manager has to consider. Portfolio managers strive to balance long-term and short-term projects and benefits, projects supporting existing markets/applications versus projects enabling new markets/applications, projects for core businesses versus new businesses, and so on. Balance is desirable across any dimension that is strategically significant for the organization.

Balance, like beauty, is in the eye of the beholder. There is no formula that dictates what the appropriate mix of projects is across any strategic dimension; rather, the answer is usually based on the nature of the organization and the experience of the executives running that organization. This is the area of portfolio management where management judgment based on experience comes to the fore. Managers use good judgment to adjust away from the strategically aligned, optimal return portfolio toward a portfolio that they believe better fulfills the total needs of the organization. Projects with a lower explicit return that build critical infrastructure or capabilities that defy easy quantification can be inserted at this stage, with an explicit recognition that this will sacrifice some short-term tangible benefits in favor of creating long-term competitive advantage that is hard to quantify.

Companies have been extremely prolific in devising a wide range of charts and graphs to display strategic balance. One of the most popular is the Project Portfolio Matrix, or risk-return grid.

This grid displays project risk (as measured by probability of success) on the vertical axis and potential value given success on the horizontal axis. The potential value is itself an expected (i.e., risk-adjusted) value, taking into consideration low, medium, and high possibilities for the most important variables determining commercial success. Projects are

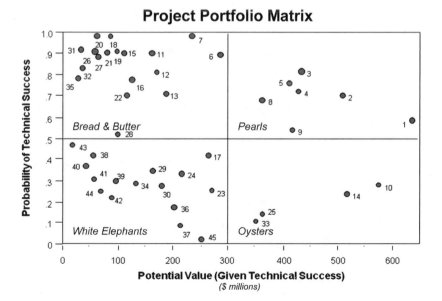

Project Portfolio Matrix

Figure 3.2

characterized as oysters, pearls, bread and butter, and white elephants. In the case of the organization displayed above, there are clearly too many white elephants, meaning projects of relatively low value with low probability of success (i.e., high risk). These resources should be redeployed if at all possible. There is also a scarcity of oysters, which represent the pearls of the future. Looking at the projects one by one had not brought this important issue into focus, where the insight leaps out from the graphical display. This portfolio was clearly out of balance. Many other displays have evolved to display and evaluate balance along other critical dimensions, such as long-term versus short-term, growth versus share, degree of maturity, degree of innovation, and so on.

Unlike return, there is no natural optimum for questions of balance. Therefore, it usually boils down to the experience, wisdom, and desires of the DRB to determine the right balance.

Making the Portfolio Doable (Living within Portfolio Constraints)

You may have identified the highest-value portfolio for any given level of financial investment, but that does not mean that the human and technical resources are available to execute all of those projects on the schedules planned. So even after a careful portfolio value optimization, we might be embarking on an undoable portfolio—one where several months down the road we may find ourselves short of one or more critical skill sets, like toxicologists, project managers, reservoir engineers, or geologists needed to execute all the projects simultaneously. Adjustments may need to be made to live within these constraints.

The final level of sophistication in assessing portfolio feasibility is to develop a resource road map for each project of the critical resource requirements out over several years to determine if some projects need to be stretched out in time to avoid resource bottlenecks down the road. In large portfolios this can be essential to avoid being blindsided by resource shortages and having the value of the portfolio eroded due to poor execution. One of the most common management failures in managing large portfolios is putting in more projects than can be executed by the available resource pool and then wondering why all the projects are missing time targets and losing value. The management levers to deal with this are reducing project resources (and therefore scope), stretching projects out (thereby delaying completion), starting projects later (also delaying completion), stopping projects temporarily, and, of course, dropping projects altogether. Some sort of portfolio road map tool is usually required to understand the implication of so many project alternatives on a large number of projects and then check the feasibility of the entire portfolio over multiple years. The result is that the DRB is presented with a set of alternative portfolios, all of which are doable with the available resources and budgets.

The Portfolio Decision

The portfolio DRB should play an active role throughout the process of portfolio analysis and optimization. In their initial review they must assess the credibility of the project evaluations, the raw material from which the portfolio analysis is constructed. In a subsequent review they must understand and approve portfolio alignment, value optimization, and strategic balance. In the final review they must understand the portfolio feasibility and approve a feasible project portfolio for funding. This is a decision of enormous complexity, for which the analysis helps build understanding and eliminate poor portfolios but cannot provide "the answer." In the end it is a decision made by the DRB informed by a lot of high-quality project evaluation and portfolio analysis. Management is still responsible for portfolio management.

The act of approving a portfolio puts the projects into two groups, approved and unapproved. This is a very primitive form of prioritization. Obviously there will be some unapproved projects that nearly made it while others missed by a mile. At this point we could revisit the efficient frontier to get a ranking of the approved and unapproved, which could be useful in case some of the approved projects encounter technical difficulties and have to be dropped from the portfolio. (This is especially important in risky portfolios like drug development, where projects can and do fail for technical reasons.) If this happens, we would like to know how to use the newly available capacity. Keeping this prioritization information up to date is a hallmark of so-called real-time portfolio management.

Building a Portfolio Analysis and Management Capability

A sound portfolio management capability requires a minimum of four elements: informed managers, involved participants, good facilitation, and appropriate process, systems, and tools. First there should be a clear portfolio governance structure in the form of a portfolio DRB. The skill of this group of managers at interpreting the project and portfolio information and making good decisions is the foundation for all that follows. There are many people who need to provide the information required to build, analyze, and optimize a portfolio. They need to know how to provide this information in an efficient, honest, and unbiased fashion, contributing to data quality. Their ability to do this is greatly enhanced by having a portfolio core team composed of skilled facilitators who know how to elicit the required information in the most unbiased way. The portfolio team also needs to know how to process and present the information so it meets the decision-making needs of the DRB. These are the first three requirements.

The fourth is having appropriate process, systems, and tools. A common failure in portfolio management is to begin by looking for portfolio systems and tools without having a process thought out, let alone the first three requirements discussed above. We want to advise strongly against the tool-driven approach. Many of the portfolio software packages available today are simply giant data repositories that can quickly and easily slice, dice, and graph the data they contain. Many of these systems pay little attention to where the data comes from and whether it is of high (or dubious) quality. We would recommend starting with a more manual process to get the first three requirements established and to test the process for its ability to provide decision-worthy information to the DRB. Once you are happy with this, you can look for the appropriate systems and tools to streamline and automate the process.

Project evaluation is an essential input to portfolio analysis. For projects that have high risk, long time frames, and/or large investments (such as drug development and oil and gas), project evaluation based on decision analysis has proved indispensable. This allows the risks to be quantified and properly factored into the portfolio optimization and strategic balance. Decision analysis also includes a number of debiasing techniques to help deliver fair comparisons from project to project. Decision analysis capability is not included in most commercial portfolio software, so usually has to be invoked separately in preparing the inputs for the portfolio systems.

In developing good project information for portfolio analysis, a three-phased approach can be useful. The first phase consists of pilot analyses of all the projects in, or contending for a place in, the portfolio. These are usually very quick, often templated analyses, requiring a few days at most. Much of the information will be gathered from the project manager and a few key project team members. Even though brief, a pilot analysis must address timing, cost, probability of success, and expected benefits over the full life cycle. For many projects this will produce data of sufficient quality to support portfolio decisions. For approximately 20 percent, additional analysis will be necessary to achieve robust data and decision quality. The next level of analysis is called prototype and often involves some market or technology information gathering, a richer financial/benefit model and more people giving inputs. Even a fraction of these will require more analysis still, especially if the project is very large, or involves multiple organizations or even companies. To implement the project, these will usually require a full investment-grade analysis, especially if a major capital investment is imminent. However, once these deeper analyses have been done, they can usually be updated fairly quickly during the next portfolio cycle.

Transition to Portfolio Capacity Optimization and Execution

We have addressed how to develop and select the "right" portfolio. We have assumed that the capacity and skill limitations are a hard constraint and that the projects must be adjusted to live within those realities. We have also assumed that if resources are not overtaxed, then the skills exist in the organization to execute the approved projects efficiently and effectively. Both of these assumptions can be relaxed. Over time, it is usually possible (although not easy) to obtain more resources in critical skill categories that are creating portfolio bottlenecks and causing good projects to be deferred or dropped. There are also many approaches, including but not limited to project/program management, to help ensure delivery excellence.

Chapter 4

Are We Optimizing Our Capacity?

John is at dinner with Bill and Chris. It's been several weeks since their first meeting, when they explored the full extent of the issues facing Acme Retail, the $6 billion consumer retail company that John joined six months before as vice president of portfolio management. After that initial dinner, they all agreed to meet up on a regular basis to address different issues. Tonight, John couldn't stop talking about resources.

John declared, "There are never enough people to go around! And if it's not people, then a key system gets taken off-line for maintenance or crashes just when one of the technical teams wants to upgrade, or one of the project managers comes and complains that the head of facilities has taken over his team's cubicles for some other group. No matter how well I think we've planned for resources across the portfolio, project managers don't stop asking for more or someone else throws us a curve ball. What do you advise portfolio managers do to optimize resources?"

Bill and Chris couldn't help smiling at each other. How many times had they heard this refrain from portfolio managers? This wasn't the first time they'd been asked this question and they knew it certainly wasn't going to be the last.

"So, what have you been doing about this?" Bill asked. He knew he wasn't being very helpful by answering a question with a question, but he also knew John well enough that he wouldn't have sat back and done nothing about it.

"Well, one of my guys was around when I was ranting about this and said that he would love to help. He's a senior manager in the IT department and was clearly itching to impress me, so I said, fine show me what you can do. I want to be able to optimize enterprise resources across our portfolio planning horizon. So I told him to report back in two weeks and we'd review what he found. Well, less than a week later the proverbial you-know-what hits the fan!"

"Wait, don't tell me," Bill held up his hand. "Let me guess. This guy ended up sending out a template to every project manager within your portfolio requiring them to list in great detail all the resources they anticipated needing over the entire lifetime of their projects. And he wanted their response by the end of the week . . . am I right?"

"How on earth did you know that? That's exactly what he did and he nearly triggered a revolution when he claimed that was what I had told him to do. So, Bill, as you seem to be so prescient on this matter, what should I have done? What would you advise we do to get a handle on resources and optimize them?"

Bill smiled and turned to Chris and suggested that she kick off the discussion about possible solutions.

"Well, John, in my experience, it's all too easy to approach this problem believing that the solution lies in controlling the detail—that by mapping every resource requirement for every project across your portfolio you'll be able to gain mastery of the situation. That's a natural urge, but it's a recipe for failure. It'll cause death by micro-management."

"Well, I could have told you that. That's why I'm in this job. It would help to have some concrete examples of what I can actually do to prevent my project managers walking out en masse." John glared at Bill, clearly frustrated.

"Look, let me put it this way, John," Bill responded in a conciliatory tone. "I've always found that what works is to focus on aggregating the detail, to focus on the big-ticket items. A good analogy is to consider your portfolio of projects in terms of boulders, rocks, pebbles, and sand. You could spend untold effort trying to 'pot' each project into one of these categories so the trick is to keep things simple. The boulders are the really big projects—strategic transformations—with multiyear timelines and impacts across the entire enterprise, and they draw on resources from all functions and departments. These are the ones that senior leadership is counting on to deliver multimillion-dollar benefits. The rocks are those projects that are less significant, maybe only touching a few departments, but still requiring significant resources to deliver major benefits. Pebbles are those projects that may involve only one department or function. And sand will be the mass of initiatives that slip under the radar. They are usually invisible except for their ability to consume vast amounts of resources—the endless requests for bug-fixes, feature enhancements, and so on. The key here is not to spend too much time worrying about which project should fit into which category—to optimize resources you only really need to look at the boulders and maybe some of the more significant rocks. The rest will take care of themselves."

John smiled. "Hey, that does make sense! I could probably make a list of my boulders and rocks right now on this napkin. But isn't that over-simplifying things? As you say, the sand tends to consume a huge amount of resource but we're effectively ignoring it by doing this . . ."

"Exactly! That's the whole point." Bill slapped the table in his excite-ment. "This is about modeling reality, but only to the point that makes sense for strategic planning. All you need to know for planning purposes are those potential resource constraints that are going to be too big for your organization to cope with in the normal course of business. The boulders and rocks are mission critical and so need to be safeguarded—the sand can always be swept under the carpet and dealt with later. Doing things this way simplifies the picture so you can do your strategic plan-ning. Chris, tell John what comes next."

Chris took a drink of water and continued. "Well, John, the resource capacity equation, as I'm sure you know, has two sides: demand and supply. Both need to be addressed to achieve resource optimization, so let's look at those now."

DEMAND-SIDE RESOURCE MANAGEMENT

John nodded, so Chris continued. "When it comes to managing the demand for resources you could easily confuse yourself with a broad array of potential resources. So, just like with classifying projects, we strongly advise portfolio managers to stick to a few. We recommend focusing on skills—the people you can use; the technology environment—the systems and/or platforms that are required for project success; and facilities—the physical space and infrastructure needed both to deliver projects and that will be impacted by project outputs."

"Okay, okay, that's great." It was clear that John was totally engaged in the conversation now. He was scribbling notes on the paper napkin. "So, just to recap, I should only worry about demand for three types of resource:

1. Skills
2. The technology environment
3. Facilities

So, how do you propose we plan for and manage skill resources?"

"John, you're as sharp as a nail," interjected Bill. "I believe that Chris was just about to tell you that!"

Chris continued as John listened with his pen poised to make more notes. "The classic resource constraint is not having enough people with the right skills or experience to deliver project outputs. As in any problem-solving exercise, the first step is to gain a clear understanding of the scale of the problem. What we recommend as the best way to assess potential resource constraints is to get your most experienced project managers to undertake a review of all boulder and rock project plans and, based on their experience, provide estimates of required skill sets over the lifetime of each project. What you're looking for is a list of key skills sets in terms of proportion of full-time employees (FTEs), the timeframe they are needed for the project and the magnitude of any identified constraint . . ."

"Hold on, you've lost me." John looked puzzled. "Give me a specific example so I know what you mean."

Bill stepped in, "Okay, so what Chris is describing is actually a two-stage process. First, get one of your trusted project managers to review a boulder project plan and make a list of all the skill types needed to deliver it, broken down into monthly blocks. There's no need to be any more granular than monthly, otherwise you'll start to get lost in the detail. Of course, the project will need a project manager, and if it's a boulder, probably also a project administrator—both for the full life of the project, say twenty-four months. Let's say this project involves all your retail stores—how many is that?"

"Oh, let me think . . . thirty-six."

"Great, so let's say that you'll need the attention of each store manager for at least one full day a month for the first nine months, rising to two full-days a month for the next six months, reaching a peak at implementation time of one week a month, and then only half a day a month for the final months of the project. You do the same for the store operations manager, retail systems analyst, and so on. That's stage one of this process and you will have lists of required people skills for each project for each month over your portfolio planning horizon.

"The next stage is when your office combines these skill estimates to get a high-level view of the portfolio-wide resource demand, by month and by skill type. This allows you to model resource hot spots across the portfolio within the planning horizon—a hot spot being defined as where there is likely to be a resource shortfall in the order of magnitude of two or three times the expected need. And, this is a key point—you may want to write this down." Bill couldn't help smiling as he tapped the paper napkin in front of John, which was already filling up with notes.

John didn't respond. He was waiting to hear the next gem from Bill. "Keep things simple by focusing only on potential resource shortfalls with an order of magnitude of at least two. Don't forget that this is a strategic planning exercise and one of the key outputs from this process is to trigger meaningful conversations with the key decision makers. Your office—the portfolio management office—doesn't have any executive decision-making powers. I like to think of your role as that of air traffic controller. You marshal the facts, present them in a logical manner, and provide options with implications so that decision makers can make *informed* decisions. We all know that in some circumstances a gut decision may be the right one to make, but only after we have reviewed all the facts. You don't see air traffic controllers working off their gut, right?"

"Hold on, Bill. Let's keep our feet on the ground here. I take your point and, no, I don't have any budgetary control over the projects within the portfolio—that belongs to the business sponsors as it should. However, what if I'm wrong? What if my so-called experienced project managers who review the resource plans mess things up? How stupid am I going to look then?"

"John, this is the whole point of the exercise. It doesn't matter if you are wrong. One of the reasons we recommend keeping these planning esti-mations at such a high level is that you avoid wasting time and energy debating silly points of detail. You're only focusing on resource constraints that are significant and that will only have a significant impact on signi-ficant projects (the boulders and rocks). You shift the debate away from whether your portfolio planning skills are good or not, and onto the real issues of how best to allocate limited resources across the organization to deliver the maximum value."

"Wow—I get it. This is fantastic!" John was smiling again.

"Great! Think of it this way. If you are facing a potential portfolio of a hundred projects and you go through this exercise—analyzing resource requirements across the enterprise and map potential critical con-straints—and you facilitate a debate within the executive team, which results in only eighty projects being given the go-ahead, but with enough of a resource contingency to significantly increase their chances of deliv-ering the expected value within the planned timeline, then you have avoided a whole world of pain—both for the business and for yourself. You know as well as I do that once projects start, people become emotionally invested in them—they become a part of their sense of success. So, if a few months down the line you pop up and declare that the business doesn't have enough resources to deliver all hundred projects, what do you think those who are invested in the twenty candidates for the chop are going to think of you? It's never an easy discussion, is it? This way you can neatly side-step the emotion and do the cutting before anyone has invested too much of their ego in it."

Bill reached under the table and fumbled with the lock on his briefcase. He clicked it open and pulled out a few sheets of paper. "Look here—I brought along a few items that I thought might come in useful tonight." He slid a sheet of paper over toward John. It looked like a page from a color presentation. "This summarizes what we've just been talking about. The point is that, at your level of the organization, you need to keep things at the aggregate level. You can keep this."

"Bill, if this was the only thing you told me about tonight, I think I'd forever be in your debt. Only today I had our senior VP of store operations pretty much spitting at me over my recommendation to cut one of his pet projects. If only we had done this earlier . . ." John looked wistfully at the sheet of paper.

Resource Management Differences

Granular (MS Resource Management)	Aggregate (Resource Constraint Analysis)
• Provides Current Information as Data is More Dynamic	• Provides Directional Information and Data is Less Dynamic
• Changes to Plans Directly Relieves Resource Constraints	• Highlights Resource Constraint Areas for Resolution

• Information is Broken Down:
  − By Task, Activity, Project
  − By Individual Resource
  − By Week, Month, Quarter, etc.
	• Information is Summarized:
	  − By All Projects
	  − By Department
	  − By Month

• Data Derived by Project Manager who Updates the Plan	• Data Derived by Resource Managers
• Typical Audience is for (but not limited to) Project Team	• Typical Audience is Resource Managers and Executives
• Rollup Management Reports	• Drill Down Is Very Limited

Figure 4.1

"So, let's run through the other resource constraint areas. Next we have the technology environment. John, are you with us?"

"Sorry, I was just thinking how I could have totally taken the wind out that SVP's sails if I had that skills analysis on my desk. Go on, I'm listening—tell me about technology constraints."

"So, another key area of potential constraint is in the ability of your technology environment—the capacity of your computer systems or platforms to cope with the demands of your project portfolio. Again, the principal here is to keep things simple, otherwise you could be lost in a blur of techno-speak. The way I recommend portfolio managers to approach this is to get the technical experts to analyze plans for the number of times a project impacts—or touches—a specific technology environment or domain in any month. It's the same as with people skills, you want to keep the timeframe in the 'per month' ballpark and simplify the technical interaction to something like 'touch.' You'll need the technology expertise to understand whether two or three touches is significant in a time period compared to, say, twenty touches, which may have little or no impact at all."

"But isn't that rather subjective?" John interjected.

"Of course it is but, as with skill constraints, what you're really seeking to achieve is an informed debate with your technology experts. Listen, John, your role at one level is no more—and no less—than a facilitator of resource allocation decisions. You don't make those decisions. What you do is present the facts and describe the implications in terms that the decision makers can understand. Sure, you should also make recommendations, but most of these are going to be pretty self-evident and those that aren't, well, those kinds of decision can only benefit from being debated in an open forum—which you provide."

"When you put it like that, my role sounds rather exciting and in the thick of things. After that meeting with the SVP of retail ops today, on top of all the hassle I was getting from the project managers, I was starting to wonder whether I really wanted to keep going. Thanks, Bill. If nothing else, you know how to spin things the right way!"

Bill smiled and looked down at the papers he had pulled from his briefcase. He plucked one from the pile and handed it over to John. "This is how a resource constraint analysis might look. It's summary data and you can easily see that information engineering resources are severely overbooked. Like I said, this gives you the necessary ammunition to go and talk with the right people."

Resource Management (example outputs)

●	Resource Name	Work	Details		Sep 21, '08							Sep 28, '08					
				F	S	S	M	T	W	T	F	S	S	M	T	W	T	F
	MSPS	0 hrs	Work															
	MSPS Advanced	192 hrs	Work	8h			8h	8h	8h									
⊟	Violet	112 hrs	Work															
	Data Dictionary	112 hrs	Work															
⊟	Kristine	1,953.2 hrs	Work	8h		0.4h	0.4h	0.4h	0.4h	0.4h			0.4h	0.4h	0.4h	0.4h	0.4h	0.4h
◆	Consolidate Admin and L	29.2 hrs	Work			0.4h	0.4h	0.4h	0.4h	0.4h			0.4h	0.4h	0.4h	0.4h	0.4h	0.4h
	Develop single source re	0 hrs	Work															
	ProSight	112 hrs	Work															
	MSPS	112 hrs	Work															
	Data Dictionary (includin	112 hrs	Work															
	Sharepoint	80 hrs	Work															
	System Maintenance	0 hrs	Work															
	System Administration	64 hrs	Work															
	New Capability (based o.	64 hrs	Work															
	Resource Pool	0 hrs	Work															

Microsoft® Project Resource Utilization

Red signifies over allocation. Project manager uses to gauge resource utilization.

ResourceGroup	Aug	Sep	Oct	Nov	Dec
Capacity Planning	146%	154%	153%	177%	108%
Information Engineering	236%	273%	213%	205%	150%
Information Security	47%	51%	61%	49%	47%
Network Engineering	118%	127%	125%	93%	104%
Collaboration	77%	64%	17%	31%	43%
Desktop Mobile	30%	48%	41%	34%	28%

Resource Constraint Management

Red signifies over allocation. Resource Managers/ERMs use to gauge resource demand vs supply.

Figure 4.2

"It's my pleasure, John. I love this stuff. Done right, it can transform the performance of an entire enterprise. Anyway, we have one more resource constraint to discuss: facilities. By this I mean physical infrastructure, networks, office space, real estate, and so on. When looking at this I break it into two components: facilities required for delivery of the project, and facilities that are going to be impacted by the output of the project. So, to illustrate, the delivery component includes office space—desks—for a project team, network connections, office support—printers, faxes, computers. Outcome facilities means, for example, real estate that is going to be acquired or divested as part of delivering a project, like a new or expanded call center. The way I suggest you analyze this is to look for blocks of greater than ten people that need facilities for more than three months. So, as an example, a new project starting in Arizona may draw in a team of fifteen people for six months, five from Denver and ten from New York. What I've found is that trying to analyze moves that are shorter than three months or for less than ten people you get lost in the details—plus, these are usually capable of being absorbed in the normal course of business."

Bill sat back and took a drink from his glass. Chris sat forward and continued. "So, John, there you have the three key demand-side resource constraints: skills, the technology environment, and facilities."

"Thanks, Chris. I understand those—but what's your advice for managing these constraints from the perspective of the portfolio office? Bill has already eulogized over meaningful discussions with the decision makers, but aren't there things I could be doing before running off and seeking decisions on cutting projects?"

"John, of course there are. As we see it, you have four levers at your control to manage capacity constraints. These are:

1. Changing timescales—you can shift projects within the portfolio to flatten resource demands over time.
2. Decouple development from roll-out—as above, this will help to flatten technical resource demand over time.
3. Descope to reduce the absolute need for resources.
4. Remove projects from the portfolio—if none of the above options are sufficient in themselves then you would consider cancelling projects to reduce resource demands to manageable proportions."

"That sounds logical to me," John said as he scribbled some more notes on the napkin. "But all we've been talking about so far is managing the demand for resources. There's a supply side too, you know. What do you have to say about that?"

SUPPLY-SIDE RESOURCE MANAGEMENT

"John, I can see that your boss was a wise man to take you on. There's not much that slips by you!" Bill leaned over and gently punched his old friend in the arm. "You are dead right. We haven't touched on managing the supply of resources—until now! I always suggest that portfolio managers . . ."

"Let me guess," cut in John. ". . . portfolio managers should keep things simple!"

"Exactly right! I suggest that you manage the supply-side using the same categories of skills, technology environment, and facilities. Let's touch on the last two first and then spend a little more time discussing skills management, since that's where you have the most potential to make a big impact across your entire organization.

"For the technology environment, you have a range of options for deploying resources in a creative way such as using an ASP model, virtualization or duplicate environments to better manage constraints. For facility supply, you can find additional space for project teams in temporary accommodations, hotels, just-in-time offices, take over a new floor in your office building, and so on."

"That all makes sense—and it's what I've been doing for one of my new projects that's just started in New Orleans. So, tell me about skill supply management." John was clearly most interested in this.

Bill continued, "Managing the supply of skills is about differentiating between your organization's core competencies—those skills and capabilities that are the essence of its competitive edge in the market—and other, more general capabilities. You really need to nurture, develop, and keep these core competencies in-house. I'll come back to these in a moment, but first let me address the other roles. All other skill-sets can be commoditized as much as possible to give you—both the portfolio and the enterprise—the ability to flex resources across the organization and into the market. By having standardized role descriptions that include specific skill requirements, by rolling out programs to cross-train your people in these so-called commoditized roles, and by enforcing best-practice documentation and audit trails in projects, you will very quickly increase your internal supply of certain categories of skill. This will also give you the option of flexing your resource needs into the market by using contractor, consulting, or partner resources."

"I like the sound of that, Bill," John said. "Give me some more detail around that."

"In my experience, the key to success with flexing into the market is in establishing and maintaining relationships with external partners so you demand resources based on your actual needs, not on an ungrounded perception that you need bodies, and you need them now. That only leads to the market taking you for a ride and your budgets being blown before you know it. Having this option to flex people, to me, is one of the keys to successful portfolio management. I don't see this as an option but an integral part of any successful portfolio people strategy. I would recommend that you select three to four partners with different competencies and geographic footprints that make the best value proposition for your business. By planning for this in advance of your actual needs you can negotiate more favorable rates so you avoid paying spot prices for outsourced labor, which are invariably higher."

"OK, that all sounds very sensible," John said. "But how do you know what resources you might need at some point in the future? Hold on, of course, I already know the answer—the demand-side skills analysis will have provided that!"

"Bravo, John! You're a quick learner."

CONCLUSION

"So, let's just recap what we've been talking about so I'm sure I've got it. This napkin and your printouts are coming home with me, that's for sure! Okay, overall, when it comes to optimizing resources across my portfolio, I need to be trying to simplify everything I'm looking at. While I must be able to model reality, I need to do it only at the level of detail that matters for strategic planning purposes. Categorizing projects as boulders, rocks, pebbles, and sand will force the focus onto the big-ticket items. Then, looking at the boulders and some of the rocks, leverage the experienced project managers to aggregate skill resource requirements to allow us to work out which skills will be constrained when—the key here is to focus only on those constraints that are of a significant order of magnitude. And we do the same for the technology environment and facilities.

"The end result is that we can then map out resource constraints across the portfolio planning horizon. Doing this will allow us to solve some of the problems by better planning but it also gives us enough information to propose solutions and make recommendations that we can then use as a basis for an open, solutions-focused discussion with the decision makers. You say that this aspect is the most critical—I can see that now—and the ultimate aim is to create an atmosphere where discussion is encouraged and compromises are reached *before* projects are initiated only to be cancelled later on. On the supply side, there are a number of creative solutions to managing resource constraints, but the key lesson I've got from our discussion is that I should be building relationships and signing framework agreements with suppliers of those skills that we know we may need, and which are not our organization's core competency. If we identify a shortage in that area, then that's a problem for the head of HR so work on."

"That's great, John. It sounds like you've really got it. There's just one more thing, before you rush off and try and do all this tomorrow. What we've been talking about is more than simply putting in place some new processes. This is a shift in the culture of the organization. Good portfolio management requires an open dialogue between different parts of a business that may not normally be used to it. You need to be aware of this and take your time in trying to bring in these changes. Take these a step at a time. Give your decision makers the time to digest what you are proposing to them. Talk to them one on one before getting them into a room together to hammer out a compromise solution. Pick a few initial 'quick wins' to demonstrate the power of this approach, to build up the confidence of your team and, possibly most important, to build up your credibility with your peers. Here's some more information you might like to take away with you. It's a summary of what we've been talking about."

Bill slid another document over to John, the EPMC Working Document on Portfolio Resource Optimization.

He continued, "I know you're smart enough to do this well. It's been a great evening talking with you about this and I never congratulated you on getting the job."

"Bill there's no need for congratulations. In fact, the dinner's on me. You and Chris have offered me way more value than I can repay over one meal. Thank you!"

EPMC WORKING DOCUMENT ON PORTFOLIO RESOURCE OPTIMIZATION

Optimizing resources is about balancing the demand for resources with the supply.

Resource Management Differences
Case Study

What

- **Granular (MS Resource Management)** is a weekly process where project managers review detailed output to identify and resolve individual resources that are over allocated for their project(s).

- **Aggregate (Resource Constraint Analysis)** is a monthly process that provides a summary view of department level resource demand vs. supply. Resource Manager and ERM input is required. While data is not dynamic, it provides directional data that highlights areas that appear to be resource constrained and investigation is required.

So What

- **Granular (MS Resource Management)** provides current and actionable data. Directly improves resource constraints as project managers can update plans to relieve specific resources that are over allocated.

- **Aggregate (Resource Constraint Analysis)** opens dialogue between Resource Managers and ERMs, and focuses discussions on areas of greatest resource constraint. Negotiating/shifting the timing and sequencing of activities helps to optimize resource capacity and minimize resource constraints.

Now What

Granular (MS Resource Management):
- Update plans more frequently so that the quality of the resource utilization data is better
- Add CBA data to MS Project so that we include long-term and planning data

Aggregate (Resource Constraint Analysis):
- Eventually MS Project reports will replace the Master Execution Plan (MEP) that ERMs use
- Solidify process to manage business resource constraints

Figure 4.3

The primary aim of resource optimization is to create an open dialogue, based on factual analysis, between the portfolio office and business project sponsors (the decision makers).

Demand-Side Resource Management

Resist trying to control the detail. Only model reality as far as it's relevant for strategic planning. Focus planning and management efforts on the "boulder" projects and some of the major "rocks."

- Boulders: strategic transformational projects that touch multiple departments, with multiyear timelines and that aim to deliver significant value to the business
- Rocks: projects that may only touch a few departments, but still require significant resources to deliver major benefits
- Pebbles: projects that may only involve one department or function and/or with in-year timelines

- Sand: all those other projects (such as bug-fixes or feature enhancements) that are usually not part of a portfolio but, unless properly managed, will consume all available resources indefinitely

For ease of planning and management, portfolio resources can be grouped into three categories:

1. Skills: availability of sufficient people with the right skills and experience
2. Technology environment: the capacity of the computer systems or platforms to cope with the demands of the portfolio
3. Facilities: physical infrastructure, networks, office space, real estate, and so on needed to deliver projects and that will be impacted by project outputs

Planning for Skills, four steps:

- Get senior project managers to review the "boulder" and "rock" project plans and use their experience to estimate required skill-sets by proportion of FTEs in monthly blocks over the project timeline
- Aggregate these skill-set FTEs in the portfolio office over the portfolio planning time horizon
- Identify resource "hot-spots" as only potential resource shortfalls with an order of magnitude of at least two
- Use this analysis to discuss options for resolution with the project sponsors and agree a solution

Planning for the Technology Environment, three steps:

- Analyze plans for the number of times a project impacts—or touches—a specific technology environment or domain in any month
- Use technical experts to advise on whether the number of "touches" in any month is going to cause issues with the system or platform
- Use this analysis to discuss options for resolution with the project sponsors and agree a solution

Planning for Facilities, three steps:

- Review in two ways:
  - Facilities needed to deliver the project (such as desks, network connections, printers, faxes, computers)

- Facilities impacted by the project outcomes (such as real estate that is going to be acquired or divested as part of delivering a project)
- Analyze facilities needs as "blocks" of ten people and for time periods of no less than three months.
- Use this analysis to discuss options for resolution with the project sponsors and agree a solution.

There are four levers to manage resource capacity constraints:

- Changing time scales: shift projects within the portfolio to flatten resource demands.
- Decouple development from roll-out to will help flatten technical resource demand.
- Descope to reduce the absolute need for resources.
- Remove projects from the portfolio: if none of the above options are sufficient then projects may have to be cancelled.

Supply-Side Resource Management

Skill supply management: differentiate between core competencies within the organization (i.e., those which give a competitive edge) and those that can be commoditized (general skill-sets not necessarily unique to the organization). Supply constraints in core competencies should be filled by training and/or recruiting FTEs from the market.

Steps to increase the in-house supply of commodity skill-sets:

- Create standard role descriptions.
- Develop best-practice documentation and processes for projects (a "body of knowledge").
- Cross-train people across the organization to increase the supply of skills.

Steps to increase the external supply of commodity skill-sets:

- Establish and maintain relationships with three to four external partners with different competencies and geographic footprints.
- Negotiate rates with these suppliers in advance of needs to avoid paying market "spot rates."
- Anticipate skill needs by using the demand-side skills analysis, and draw on suppliers using agreed rates.

Technology environment supply management: consider using ASP model, virtualization, or duplicate environments to better manage constraints.

Facilities supply management: consider creative solutions such as using temporary accommodation, hotels, Regus offices, or taking over a new floor in the office building.

Final point to remember: when seeking to implement these for the first time, focus on a staged approach, using quick wins to build momentum and buy-in.

Chapter 5

How Well Are We Executing?

The day after the dinner, Christine Jones (Chris to her friends) thought back to her start in PPM.

She remembered when she was the lead of Project Portfolio Management for the company, but at the time she did not even know she was the lead. She had been hired for a different role and was busy learning about the company. But part of her new job involved administration of the company's governance council that evaluated, approved, and monitored technology projects. It was a council made up of a combination of technology and business leaders, and it gathered once a month.

Chris was learning about the council, how it operated, and what it required to perform the role it filled at the company. But at the time Chris didn't know that what she was doing was actually "Project Portfolio Management" or that there was a field of study that went with it. She only knew that the project governance council was performing this role and that she was charged with assisting its activity.

That changed one winter morning when she attended a quarterly meeting of the executive planning committee made up of the most senior executives at the company including the CEO. Attending the meeting is a relative term in that she was far from the large U-shaped table the executives were sitting around. Chris's boss, the CIO, in her role as lead of the project governance council, was presenting the current status of the portfolio of projects to the executive planning committee. Chris had assisted in creating this presentation and was in the back of the room nervously waiting to see how it went. The presentation was going just fine, but the questions that came after changed the way Chris approached her role with the project governance council.

The presentation was an update of the status of some of the large technology projects, how they were progressing and the forecasted spend on projects that year. When it ended the CEO asked the first question. "Since we ask that projects come to the council with a cost estimate accuracy of plus or minus 10 percent, how often do they finish within that range? He had another: "Do all of the technology projects support one of our current corporate priorities and, if so, what percent of the total project spend is directed toward each priority?" He continued, "What is the prioritization for the entire portfolio and what do the business leaders agree are the top ten projects?" And he finished with a fourth question: "Since all of the projects are approved with a business case involving projected benefits, how often do we achieve those benefits and to what level?"

Chris knew that these were great questions that currently could not be answered. The CIO did well to explain what the council currently knew and what would need more investigation, but Chris quickly wrote down the CEO's four questions and asked herself some questions. "How will we get to the point of being able these questions? How will we determine how well we are executing Project Portfolio Management now, and how will we develop the road map to be able to answer these questions in the future?"

The first thing Chris needed to do was create an overall visual representation of where the organization currently stood. She needed to create a process map of the current portfolio management activity at the company (see Figure 5.1). How do projects start? How does the company track projects early on? What are the first approval steps? She needed to follow this line of questioning through the various lifecycle steps that were currently being completed. What happens when a project ends? How is it evaluated? How do we determine if the forecasted benefits were achieved?

Chris put all of the steps into a process workflow. She thought to herself, "At this point, put the process in as it stands now. Only put the steps in that we have in place currently. Keep all the steps in the process even if we do not complete each step perfectly now." She wanted the current reality on paper, so she painstakingly drew a process diagram that looked something like this:

Next came the fun part for Chris. She added the steps that were not currently in the process, but would be needed if she were going to be able to answer the CEO's questions from the planning meeting. She added a step

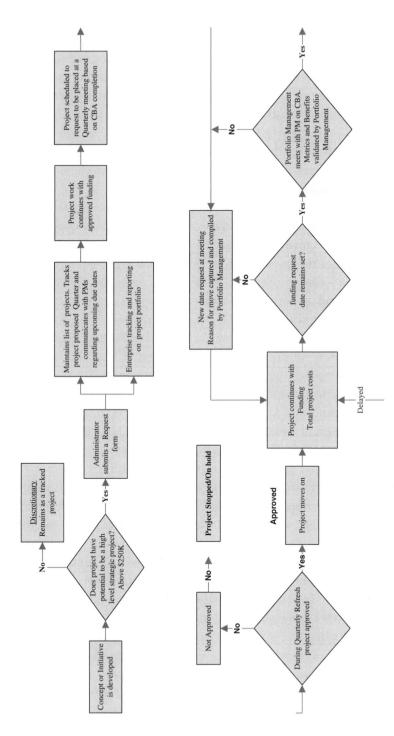

Figure 5.1 This PPM process map depicts the steps taken throughout the project portfolio management lifecycle.

for when baseline metrics are determined for projects. Since project end dates were constantly changing and project managers currently asked for governance approval when a later end date was needed, it was difficult to say at the end of the project, "Did the project take longer than its approved time and, if so, by how much?" Accounting for the fact that project end dates would change during the life of the project, especially when the project was in the very early stages, Chris established a time in the funding lifecycle when the baseline ending date would be set and all further "on-time" metrics for the project would be based on the end date established at that time.

Chris also added a requirement that all projects determine if the goals for the project align to one of the top corporate priorities. This step could be completed very early on in the project lifecycle since it was part of the high-level benefits of doing the project. By completing this step, Chris knew she would be able to correlate all the projects to their respective corporate priority and create a pie chart showing which corporate priorities were being supported and to what extent. Chris knew this would be important since the CEO had asked for it. But it wasn't just because the CEO wanted to know what priorities were being funded, but that he would want to direct those percentages as the priorities changed.

The next steps Chris added that were not part of the current process flow were the steps needed to prioritize all the projects in the portfolio. Currently, once projects were approved, they were added to the list of funded projects underway. Chris inserted the steps necessary in the process to prioritize the projects. This meant determining when each project would receive its relative prioritized ranking and how that would be tracked. While the whole subprocess for creating and maintaining a prioritized list would need to come next, Chris could insert the process into her overall process flow now.

Finally, Chris created the process flow for what would happen at the end of the projects. Currently, the projects' end dates were somewhat vague. The projects all had an end date, but even projects that were finishing on time were somewhat loose about whether or not the project was in fact "done." Chris had to define what "done" meant and then lay out the process for what happens after that. When does a project come back to the governance board to report on the relative success of the project success (i.e., the on-time and on-budget type metrics)? This would also be the time that Chris would want the project managers to discuss lessons learned and best practices, so Chris added that in.

There would also need to be a specific process and timeline for assessing the extent of benefit realization. Chris added to the process flow a series of steps one year after the project had completed. What would the project need to do at this point? What role would the business sponsor play in determining and assessing project benefit attainment?

It was during this section that Chris started to understand an added benefit of the process flow diagram. Since she was adding a section into the workflow on benefit attainment, she knew she would need to address another issue the company was facing. The business cases presented during funding request were inconsistent and at times lacked good project metrics. If the original business case was vague as to how benefits would be measured, then it was going to be very difficult to assess whether the project realized those benefits at the end. Chris went back to that section of the workflow and added some steps to ensure that there was a business case review process for quality and consistency before the actual review at the governance board.

There was one last step that Chris knew she had to add. She had to look at the process steps and honestly assess the extent to which they were being completed at her company. For this she gathered a few people at the company familiar with the process and pulled out the yellow highlighter. "Let's highlight the boxes we currently don't do or are completing inconsistently," she told the small group. Box by box, step by step, Chris and her coworkers highlighted some boxes and left others intact. When they were done, Chris stepped back. The diagram before her showed a clear picture. There were many highlighted boxes in the beginning of the process map and many at the end, indicating those steps needed improvement. The middle section was relatively clear.

Chris could now answer the question "How well are we executing?" Not only that, she could see where to start to fix the problems. She could develop a strategy to improve the steps that were highlighted. She could determine which ones to start with and which ones would need to wait. She could also mark success along the way and show her progress to executives who were asking, "Are we getting any better at this?" They would be getting better and Chris would be able to show them where they were, how they were getting better, and what was next.

Chapter 6

Can We Absorb All the Changes?

John Durham has been in his role of vice president of portfolio management at Acme Retail Inc. for almost nine months and he is wrestling with a problem. It's something that's been bubbling just under the surface for a while, but recently has emerged as a hot issue that can no longer be ignored. Several key projects had just delivered, as planned, within the past three weeks, but reports were pouring in of low employee morale, increased attrition rate, growing complaints, and a noticeable drop in productivity. The Chief Operations Officer, Leo Palmer, had come to see John the day before and wanted to know what was happening and what they should do to turn things round.

"John, I can't put my finger on it," he said, "but I just think people are sick of all this change. We've all grown up in a business world where the mantra has been 'change or die' and Hannah has made it clear to everyone from the moment she took over as CEO that she was going to shake things up. Tell me if I'm mad, but I really think that we've gone too far too fast."

"Leo, I haven't been around that long, but I have to say that I was thinking just the same thing myself only yesterday when I was reviewing our project performance reports. We're just not delivering as well as we were and it was when I took a step back to look at the big picture that it occurred to me that these three projects that have just moved into implementation phase are a major cause of all this angst."

"So, tell me, John, do you have a nice little chart I can take to Hannah Choi that will clearly show we need to slow things down a little? I'm going to need more than a general assertion that our people are reeling from all these initiatives. Without hard data, she'll bite my head off and hand it back to me on a platter! You're our project portfolio management guru . . . can you show me how much change is too much?"

John had told Leo that, of course, he could come up with the necessary data. After all, what was the point of having the portfolio management office if it couldn't serve the business and deliver real value?

That was yesterday, and now John was sitting at his desk feeling very uncomfortable. He had some ideas on how to tackle this challenge, but he had to admit that he really wasn't sure how to work out how much change was too much for Acme Retail Inc. It seemed obvious to him that those three major projects must be part of the problem, but something else was telling him that they might not be the whole problem. Furthermore, they were critical projects for the business transformation that Hannah had made clear was her top priority. There was no way that he was going to allow any delays to those initiatives when there were so many other projects that could be shifted to ease the load on the workforce. He needed a way to assess change across the whole portfolio. His hand hovered above the phone; he hesitated, then sighed and dialed a number.

"Hello, Bill, it's me, John. Hi, yes things are fine. . . . Well, actually, I have a problem. Do you have a few minutes to chat?" John explained the problem. "Okay, that sounds great." John turned to his computer and checked his calendar. "Great, let's do it in an hour. Bill, you are a star—I'll owe you big time on this!"

DEFINING CHANGE

An hour later John closed the door to his office, put his phone on speaker, and dialed a number. Bill's voice filled the small room.

Bill started to talk. "So you need to know how much change is too much for your organization, right? Presumably you're also wondering what to do about it once you've worked out what's going on. Well, John, you've come to the right place." Even over the phone, John could detect the grin that was creasing Bill's face.

"Bill, you're the man! So, let me tell you what's really bugging me right now. How can I measure change? How can I get a handle on what impact all these projects are having—and are going to have— on the organization? I mean, it feels like too much change, but how can I demonstrate it to Hannah Choi, our CEO, who doesn't like anything that doesn't have hard numbers to back it up?"

Bill responded, "As in any problem-solving exercise, the first thing is to get your arms around the problem itself. It will help to define all this change in simple terms, remembering the principle that when managing a portfolio, all you need to model reality is just enough information to satisfy the needs of strategic planning. Do not allow yourself to become sucked down into the detail."

"There's change that impacts technology, there's change that impacts physical assets (such as real estate), and then there are the changes that impact people. It's this last category that matters—it's only people who really get unsettled by change. Technology and physical assets don't have emotions. So, the focus of your work needs to be on the types of change you're driving onto people, and over what timeframe. There's clearly a world of difference for people undergoing change once a year compared to once a month."

"What I'm going to outline to you now is a fact-based methodology that will get you to look at change in terms of *what* (the degree of disruption), *when* (the timing of these changes), and *who* is impacted (both individuals and groups of people). You will then be able to present your analysis, along with recommendations, to your CEO so she can make the necessary decisions to minimize adverse impacts and maximize the value delivered across the organization."

"Furthermore, John, once you've started this process and controlled the impact of change across your business, you'll be able to slowly, controllably increase the capacity of the entire organization to handle more change."

"Great!" said John. "That's what'll keep Hannah happy. Maybe I'll have to raise an issue she'd rather not hear, but at least I can give her some positive news too. So, where do we start, Bill?"

TYPES OF CHANGE

"John, tell me one thing. How many projects are you running right now in your portfolio?"

Without even pausing to check his numbers, John fired back, "Ninety-seven projects within nineteen portfolio programs, and eighteen divisional programs."

"Okay, so there's a lot going on. Before we delve into these projects, let's just take a step back and think about change and how people cope with it. I presume you'd agree that there are different types of change?"

"Sure—it goes without saying that the change involved in using an upgraded version of, say, Microsoft Office is nothing compared to what someone has to go through when they're required to stop performing a task manually and instead use a new ERP system."

"Exactly," replied Bill. "So, what we need to do is classify the type of change. When doing this I recommend four categories, each with an escalating impact on your people. The change with lowest impact is what I call Affiliation Change. Basically, this means a change in someone's relationship—whether organizational, like a change to their reporting structure, or physical, like a change in their location. These are relatively low impact (of course, there are exceptions, but usually these are as a result of a combination with the other change types)."

John was making notes. "Got it—affiliation change: low impact."

"Next in terms of impact comes what I call task change. This is any change that relates to the way someone does their work and could include new a process or activity, like changing the way someone updates information in a computer system, or a new way of accessing customer information. After that, you have skill change, which requires people to learn new skills or acquire new capabilities. An example here could be a task that has become computerized so the operator has to learn new skills to interact properly with the computer. Then we have the change with the greatest impact, which I call belief change . . ."

"*Belief* change? We're not asking anyone to change religion here!" John smiled, unable to stop himself from interrupting.

"Well, if you'd let me explain, I'll tell you. This is the most difficult form of change to handle because it involves changing the way people think about the work they do—it requires a new attitude, a new approach. An example of this could be from implementing a new way of managing a call center. Previously, operators may have been incentivized on the number of calls they handled in an hour. You can see how this would build an approach to calls where the operators would either try to deal with the caller immediately or pass them on to the relevant technical expert as quickly as possible. If the new way of managing calls was to build satisfaction by having customer problems solved as much as possible on their first call, the operators would have to change their approach to try and close as many of the calls themselves, and only pass them on to the technical experts after they had exhausted all other avenues."

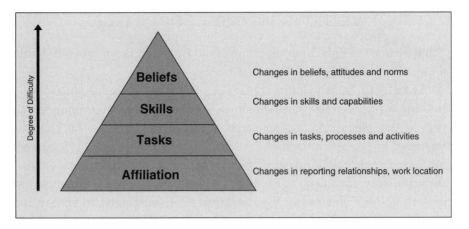

Figure 6.1

"This all makes sense, Bill. What you've described sounds like a hierarchy of change impact."

"Exactly right—and I usually illustrate it as a four-layered pyramid with affiliation change at the bottom and belief change at the apex. Let me email you a picture of what I'm talking about."

Quickly checking the picture in his email, John said, "But this doesn't really cover all the things that people have to deal with on a daily basis. It seems a little simplistic just to categorize project-induced change. I mean, take last week, for example. Despite trying to shepherd three critical projects through their rollout stage, I had to spend a whole chunk of time filling in our half-year appraisal forms—we have a very rigorous process for assessing staff, which is great, until you have to complete loads of 360-degree review forms within a week! How would you classify that kind of activity?"

"You've hit on a very important point, John. Project-induced change is not everything. When considering the impact of change on your people you also need to factor in all the other standard enterprise activities that they're involved in, just like your annual appraisal process, or, for your accounting department, the budgeting cycle or closing the monthly, quarterly, and annual books. These are equally important in assessing the impact of change as they absorb mental and emotional bandwidth. What we see is that people become less able to cope with change at times when they have lots of other demands on their capacity. So it's just as important to capture these kinds of activities and factor them into the model."

MODELING THE IMPACT OF CHANGE

"That brings me right to my next question, Bill. How do you turn this into a model that I can analyze and control?

"The way to start modeling the impact of change on your organization is to bring together your project business sponsors and project managers to review the plans and assess three things. First, assess the changes triggered by projects in terms of the four dimensions we've just discussed: affiliation, task, skill, and belief. Next, you'll need to assess timing—when these impacts affect people. For this, I recommend you place the impact at project 'go-live.' For really big initiatives, you may need to spread the impact over several time periods, but I wouldn't recommend getting any more granular than monthly. Finally, you'll need to articulate who is affected by these changes. This can be broken down into categories— something like a split between executives, managers, and employees works well—as well as by department or functional area, like accounting or sales."

"Okay, this sounds very straightforward, Bill, but do you really do this in one hit like you describe? I sense that this is not a simple linear process—there are too many variables at play to be able to do this in a mechanistic way." John was frowning.

"No, you're absolutely right. What I've described is just the start of the process for achieving a robust consensus view of the change profile across your enterprise. What usually happens—and if this doesn't, I would strongly advise you to make this happen—is that you would do an initial run of your change model and then reengage with key business sponsors and managers to validate and clarify the results . . ."

"Hold on there a second. When you say model, what do you mean? Have you got a piece of software that does all the number crunching for you? What is it, so I can get a copy?"

"John, this can all be done in a spreadsheet. All you need to do is create tables to input your data, process the numbers, and produce charts to illustrate the outputs. The issue isn't so much about running the data, it's about making sure that your inputs and assumptions are valid. Look, at the end of the day, all this tool does is facilitate an objective discussion between you—the portfolio management office—and the business leadership. As in most management problems, the solution usually lies at the end of an open discussion based on factual analysis and mixed in with some good old-fashioned management insight."

"Bill, of course, you're right on that. I get that completely. I spend a lot of my time walking the corridors of power. I just wanted to check whether this was going to cost me some more software licenses, but we've got plenty of 'spreadsheet jockeys.' Sorry I interrupted you. You were telling me about clarifying the results of the initial model run."

"Yes, once you've got your initial results, you should take them back to the project managers and business sponsors to check that they agree with the outputs. What you're also looking for at this stage is to aggregate any overlap in project impacts. It's quite possible that several projects will be delivering what is effectively the same change to the people concerned, so you don't want to double-count these in your outputs."

"Bill, sorry to interrupt again, but when you say outputs, what exactly do you mean? What do these look like?"

"That's another good question, John. The outputs are usually expressed in graphical form with time along the bottom axis—like I said, try and keep it to a monthly scale—and change impact on the vertical scale. And before you ask, let me explain a little about the functioning of this metric. The change impact metric is a function of the type of change (affiliation, task, skill, or belief) and the number of people affected. You can apply different weightings to each change type to reflect the increasing impact each has on your people. So affiliation change could have a weighting of one, task change could be weighted as two, skill change could be three times, and belief change could have a weighting of four or even five. The actual output is a number between one and a hundred and represents the percent change that people are experiencing.

"Put all this together and you can 'slice and dice' the data to present graphical illustrations of the timeline of impact across the enterprise by department, by change type, by portfolio, and so on. Very quickly you've created a powerful visual image from which anyone can draw the necessary conclusions. Here, let me send you another email with a picture of what I'm talking about."

"Bill, that sounds perfect! I knew I could count on you to come up with something that would help. Hannah is very much into facts in numbers, and she likes them presented in a visual format. But what about the underlying business activities you mentioned earlier? How are they factored into this model?"

"Well, every business has periods in the year when people are busier than at other times. As I've said, change is easier to absorb when people are close to their operating 'steady state' and are not stressed by unusual

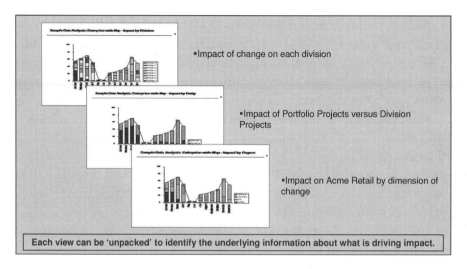

Figure 6.2

pressures, such as budget time in your accounting department. So what you need to do as part of the process of engaging with the business is to ask each division or function to identify the activities and events that happen on a predictable pattern that take their time and attention. These will be the times when their bandwidth to absorb new ideas, learn new tasks, develop new skills, or think differently about their work will be most limited. In essence, what you'll be doing when you complete this aspect of the model is revealing the underlying business cycle for your enterprise. But the real win for you is that this will give you the understanding of how the capacity of your organization to cope with change evolves over time."

"Bill, that's amazing. In a matter of minutes, you've explained something to me that could revolutionize the impact we, in the portfolio management office, can have on the business. This is incredibly powerful stuff!"

CONTROLLING THE IMPACT OF CHANGE

"John, I'm glad you find this useful. But I'm not quite finished yet. Clearly, once you present your findings to your CEO, she'll look to you for recommendations on ways to reduce the impact of all the change on the organization."

"Yes, you're right there. But I think the answers are pretty obvious, aren't they? Rescheduling project rollout dates, rescoping projects, or minimizing the groups who are impacted by any changes are things that come to mind immediately. Oh, and cancelling them completely, though I imagine that's not a popular option!"

Bill laughed. "Yes, you could say that, though what this approach does allow you to do is propose the cancellation of a small, low-priority project that just happens to have big change impact. What you'll find is that it's not always the big high-priority projects that cause the most 'change stress' in an organization—which is a very good thing, I hasten to add.

"But let's just take a step back for a moment. You asked me up front how you could work out just how much change is acceptable for an organization. To answer that, I tend to follow industry best-practice recommendations, which can be summarized as follows: Anything up to 10 percent change is experienced by people as good. They feel that their organization is not static and is moving with, or just ahead of, the market. Change of 10 to 25 percent change feels challenging—people feel that their organization is undergoing significant change and this will be challenging to manage. Change of 25 to 50 percent feels very challenging to people, and this will be very stressful to manage and sustain for any length of time. In terms of this model, this is the 'red' or 'caution' zone. Anything more than 50 percent change is the danger zone where people are being asked to make major changes in what they do and how they identify themselves— this is where you'll need to take immediate action to bring the change back under control. So, the first thing you need to do is prioritize your change-control actions so that you can reduce any immediate problems. However, the real value comes in anticipating where the impact of change could strike next and plan to avoid this completely."

"Fine, that all makes sense. And when you make me look at it like that, I can now see a whole range of other actions that we can take to mitigate the impact of all this change. For example, presumably we could introduce earlier communication and education on some of the bigger changes that we are planning. Or we could stagger some of the rollout across departments to coincide with months when they are least burdened by their regular activities."

Bill chuckled out loud. "It sounds like my work is done . . . now you're thinking along the right lines. What I find is that this model generates a load of ideas that you and the business leadership team may not have come up with on your own without visibility of the change impact constraints. But the real long-term value comes from the very fact that you are having these discussions at all."

CONCLUSION

Bill continued and the tone of his voice lowered as if to emphasize the importance of what he was about to say. "John, what I've just described to you is more than just a methodology for calculating the impact of change on an organization. It's a tool that will allow you, and the senior management team, to build a more resilient organization, one that's better able to absorb change and that, over time, increases its capacity to change—as your people see that they can cope with change, so their capacity to handle it will grow. You'll be in a position to predict with greater accuracy which projects are going to succeed because you're planning for success from the start. And if you start to see potential problems looming on your planning horizon, you'll be able to react in time to prevent major issues, or at least mitigate their effects. This is what portfolio management is all about. Let me just email you one more thing—an EPMC working document on Enterprise Change Management that sort of sums up what we've been talking about."

EPMC WORKING DOCUMENT ON ENTERPRISE CHANGE MANAGEMENT

Enterprise Change Management is about providing visibility and understanding on what is the "impact of projects and change" on people, workgroups, and departments within an organization.

Enterprise Change Management answers the following questions:

- What is our capacity for change?
- How much is too much?
- How should we priorities and focus our efforts? and
- What decisions need to be made to improve the probability of our success?

An Enterprise Change Management Assessment Tool can be used to measure and assess the overall "ability to absorb" the impact and change across individuals, groups, and departments within the enterprise. By identifying impact points and the subsequent capacity for change across multiple dimensions, a framework for decision making and prioritizing of projects—along with the associated resources, project execution, and scheduling—can be accomplished. It is important to remember this is not a one-time assessment but a tool that can be used by the organization on an ongoing basis.

The organizational impact assessment is a fact-based assessment and view of what, when, and who is impacted. The assessment delivers executive decision-making capabilities around priorities and scheduling so that projects are delivered as promised. The assessment is straightforward and has three key impact focus areas:

- Expected timing of activity/projects
- Number and involvement of resources
- Degree of disruption

If done properly the output will:

- Be an inventory or catalog of all the organization's "change impacts"—who is impacted, to what extent, and when
- Have a view of capacity, readiness, and alignment for change across the organization by workgroup, department, and enterprise
- Be a simple database that allows an organization to view "impacts" on a rolling 24-month calendar
- Include identified and prioritized impacts and change risks across each project via "heat mapping"
- Have recommendations to keep an organization from overextending themselves and allowing too much change within the enterprise

Any company that has more than a handful of projects should have this type of tool in their toolbox.

Time and again, the research shows the importance of matching change impact to change capacity and the drive to understand the impact of change at an individual level is needed to see the big picture.

Chapter 7

Are We Realizing the Promised Benefits?

"Does it getting any better?" wondered Bill. "An assignment completed ahead of time, a happy customer, and the opportunity to get home early for the weekend." Then he made the fatal mistake—he called in to the office just to check up on things and pick up the mail.

Bill's office assistant knocked on the door, "Sir, there was a call from John Durham. He said it wasn't urgent, but he'd like you to call back as soon as possible."

"Thanks. Acme Retail," thought Bill, "I seem to remember saying my work there was done—looks like I might have been a little premature!" And then he heard his mother's voice in his head: "Never put off until tomorrow what you can do today." Bill sighed and reached for the receiver—he never ignored his mother.

"Hey, Bill, thanks for calling back. I just wanted to say thanks for all your help. We're really making progress in sorting the issues we talked about. I can honestly say we're aligning our investments with our corporate strategy and managing project delivery and our capacity more effectively. Leo and Hannah are really happy."

"That's great," replied Bill. "So why the call—what's the problem?"

"There's no problem as such—or at least I don't think there is. It's just that I saw Hannah in the elevator this morning. She commented on the progress we're making and said how impressed she was by the last portfolio report. But then she said something that has been nagging at me ever since . . ."

"Let me guess—something about she can't wait to see the results in improved benefits realization across the portfolio."

"How'd you know . . . don't answer that—you always seem to anticipate the next question."

"Thanks. I'll take that as a compliment, but I can't claim clairvoyance on this occasion. I read the financial pages and saw that Hannah was getting a grilling from the analysts on when she thought she'd be able to show a return on the investment in her new strategy. So what's your problem?"

"Well, it's as I said, we've made real progress in terms of selecting projects and managing their delivery, which should translate into benefits, but I've got this nagging doubt because we've never been asked to demonstrate the benefits before—I kind of always thought they just happened, but I'm not so sure now."

"Don't worry," said Bill in as reassuring a tone as he could muster on a Friday afternoon. "The progress you've made to date has laid the basis for benefits realization. But you're right to be concerned because there's one other issue we've not covered in our chats to date."

"What's that?" asked John.

"The starting point is to realize that benefits management is not an add-on to the portfolio management process—rather it's an attitude and focus that runs throughout everything we do. We invest in projects and programs to achieve benefits in terms of increased revenue or sales, cost and time efficiency savings, adherence to regulatory and legal requirements, maintenance of business as usual, contributions to strategic priorities, and to achieve business performance improvements."

"I know that!" exclaimed John. "We're taught that on day one of our PPM training. It's just translating that into the real world seems a little more . . . problematic."

"You're right, and I've a couple of techniques that can help you 'lift the fog,' but first it's crucial that we recognize that the problems you're facing are common to many and as such represent a puzzle—we all say we invest to realize benefits, but most of us can't show the benefits of our investments! It's a paradox, and the implications are serious. It isn't just about problems in demonstrating benefits—it goes way beyond that. Look, if we're not sure about benefits, then how do we appraise our investments, prioritize the portfolio, or ensure that we're optimizing the return from our accumulated investment in change?"

"When you put it like that I don't reckon there's much argument—but how do we resolve this problem?"

"As always, the practitioner cuts to the chase. Well, there are three keys to the benefits management paradox or puzzle—let's talk about them in turn."

KEY 1: ENSURING ALL BENEFITS CLAIMED ARE ROBUST AND REALIZABLE

"First, we need to ensure that all benefits forecast in business cases are robust and realizable," Bill continued. "This provides a degree of confidence in the claims made and helps ensure we select the right projects for our portfolio."

"That sounds great, but how do we go about that in practice?" asked John.

"Well the problem that we find in many organizations is that the business case overstates the benefits to get funding, but that means we don't have a reliable basis on which to prioritize our available funds and then, in due course, we don't actually realize those benefits."

"Tell me about it," interrupted John. "I feel kind of 'poacher turned game keeper' here—I used to write business cases in my last job and most of the time the task was to present the case in the most favorable terms. Not lying, you understand, it's just that we wore some very rose-tinted spectacles at times."

"Don't feel too bad about it," Bill reassured John. "It's not uncommon, and there are three things you need to consider in dealing with the problem."

"First, establish a set of rules about how benefits will be classified, quantified, and valued. When we consider costs, our friends in finance ensure that we use a consistent rule set, but when it comes to the other side of the value for money coin—benefits—too often anything goes. So step number one is to ensure you have a rule set that applies to benefits across the portfolio, ensuring a level playing field for portfolio selection and prioritization decisions, and for tracking benefits through to realization. The key is that whenever we make an investment, we should be clear about the benefits that we are buying—what they are, when they will be realized, and how we'll know that they have been realized. These questions should be formally answered in a Benefits Realization Plan for each significant investment."

"Second, we need to validate claims with an independent review by the PMO and by agreeing to the benefits with the recipients. This is important for two reasons—first, while projects deliver capability, it's the business that realizes the benefits, and second, they are often realized after the project has been shut down. But there's one proviso: this agreement needs to occur before we make the investment. In practice, we find that getting people to sign up to realize benefits after the investment has been made is . . . what term did you use earlier . . . problematic?"

"Ok, I buy that," replied John. "But when you say 'agree to them with the recipient,' that sounds great in principle, but who do you mean in practice?"

"Well, it varies with the benefit and the organization. But the idea is anyone who has some responsibility for the performance of the business unit that will benefit from the project. So that could mean the head of the Strategic Business Unit, or the business change manager, or the strategic planning function or even the finance director."

"Okay . . . sorry, I interrupted you. What's the third thing we need to consider?"

"Thanks," Bill continued. "It's that we have regular checks that the benefit forecasts are still reasonable. The good news is that you've already got the process in place for this—in your stage gate and regular portfolio reviews. All you need to do is ensure that these reviews focus on asking whether the benefits forecast are still realizable and whether the business recipients agree to realize them. That takes us to our second key— ensuring all forms of potential value are captured. But does what I've been saying make sense so far?"

"Absolutely. I've found an unused corner of that napkin from our first meeting and am scribbling away. I keep it here in the office, as a reminder of the great advice you gave me."

"Each to his own, and if it works for you, John, who am I to criticize?" laughed Bill. "Okay, on to the second key."

KEY 2: CAPTURING ALL VALUE CREATED

"Too often projects hit the hurdle rate and then stop," Bill continued, "but that means potential benefits may be ignored and consequently lost, since the business change required to realise them won't occur. The answer is to ensure we look at all investments from the other end of the kaleidoscope— rather than asking are there sufficient benefits to justify the investment, we should be asking, Have all potential benefits from this investment been identified?"

"That's great—I presume you have a framework that I can use?" John asked hopefully.

"Of course, grasshopper, listen and you shall be enlightened . . . capturing all potential value means that we need to focus on four main categories of benefit."

Bill picked up on the "grasshopper" comment and recalled that in college, Bill had been a big fan of the television series *Kung Fu*. It appeared he was still a fan. "Maybe it's all the reruns of the show now on the cable channels," Bill thought to himself.

"First, efficiency benefits, which include both time and financial savings. The key here is to borrow from Steve McGarrett."

"What?" For once John was completely lost.

"You remember, the old television series *Hawaii Five-O* and the immortal line McGarrett always uttered when they finally caught up with the villain: "Book 'em, Danno!""

"Oh, yes, I remember. I think my dad told me about it."

"Don't forget—we *are* the same age," Bill retorted.

"Ouch!" exclaimed John. "Touché!"

"Now, back to what I was saying," said Bill. "The advice is sound—book the benefits in unit budgets and headcount targets. Although you also need to check that the benefit is real."

"What do you mean?" asked John. "Surely if you've booked the benefit in a budget cut then it's been realized?"

"Not necessarily—you can cut the budget for sure, but if the forecast efficiencies are not realized, then all you will really have achieved is an unfunded pressure and your budget cut will be at the expense of output or service quality."

"Oh, I see—you're right. Is there anything else to look out for?" asked John.

"The other fundamental thing to watch out for is time savings. The point here is that okay, an activity might be completed more quickly, but benefit is only realized when the time saved is used for some value-adding activity. Until then the benefit is only potential. Look—do you still shop at Safeway?" asked Bill.

"Yes, but I don't see what that's got to do with time savings," replied a rather confused John.

"Well think of time savings as a voucher—I've got a $10 Safeway voucher. Do you want to buy it off me?"

"No, no thanks."

"Why not? You already admitted you shop at Safeway, so give me $10 and you'll have a shiny voucher for $10 to use this weekend."

"But I might not go shopping!" objected John, wondering where this was going.

"Exactly—and that's the point about time savings. The time saved may not be realized for some time by simply redeploying to value-adding activity, particularly where it's five minutes here and five minutes there. So watch out for business cases that value such benefits as the cost of labor, assuming that 100 percent is converted to value-adding activities as soon as the system is implemented. It just doesn't happen in practice."

"Okay—thanks. Although I'm sure I'm not going to be too popular with our business case writers."

"Maybe not, but you don't work for them. Think about the shareholders! Anyway, that's enough on efficiency benefits. The second main category concerns effectiveness benefits, in terms of improved performance and contribution to strategic priorities."

"I think we're ahead of you here, Bill. We've introduced a process for assessing strategic fit by rating all projects against our key strategies with a score of five for mission critical projects, three for those with a major contribution, and one for those with only a minor impact on the strategy," interjected John proudly—for once he was one step ahead of his old friend.

"That's encouraging, but I'm not talking about strategic alignment or fit, but strategic contribution. What you've described is a step in the right direction, but it doesn't tell me anything about how the project will contribute to the strategy, when, or by how much. The key here is to ensure you understand exactly how a project will contribute to achievement of a strategy and how that impact will be measured. A useful technique is to use what we call Strategic Contribution Analysis, which combines strategy mapping (mapping from vision through strategy to success measures) with benefits mapping, which shows what benefits will result from an initiative and any business change on which realization of those benefits is dependent. By combining the two we get a clear line of sight from a project through to the organization's measures of strategic success."

"Oh, I see what you mean," said John, only partly hiding his embarrassment at his too early signalling of victory.

Bill continued, "Our third main category of benefits relates to investments in infrastructure. The problem here is that the full value of such investments lies in the flexibility that they provide for us to support both existing and new applications and users, but it's not immediately obvious how we can value this flexibility."

"Yes, we've struggled here," said John. "We had a major infrastructure business case for review only a couple of weeks ago. Please don't tell me you're recommending real options theory, because that's what the business sponsor had done. He'd recruited some hot-shot consultant who produced a business case using real options, but no one, including the business sponsor as it turned out, could understand what it was saying. The investment committee sent them away to convert the case into English that we could all understand."

"That's not uncommon, and I think it confirms that Acme is making real strides—to refuse to say yes to an investment because management doesn't understand the value it will receive indicates that you've made some great strides—well done, John."

Bill was aware that his old friend's confidence was in need of some replenishment. "To reassure you, I'm not talking about real options theory; rather, I'm suggesting we use what we call a Potential Opportunity Value approach in which a proportion of the value of the portfolio pipeline is taken into account based on an assessment of the probability that the projects that will deliver these benefits will come to fruition. This is not as difficult as it sounds—I promise! One of the by-products of implementing portfolio management is that you get a single view of the organization's project pipeline of actual and planned projects. This, along with the organization's applications road map, can be reviewed to identify all planned applications that will depend on the new infrastructure. We then adjust the total value of these applications by a factor to reflect the fact that not all planned applications will get approval. Clearly the choice of the probability factor will be somewhat subjective, but a degree of objectivity can be obtained by evaluating the potential applications against criteria such as scale of stakeholder commitment; whether it meets a legal or regulatory requirement; the availability of funding for the project; current stage of project approval (i.e., how far through the investment pipeline/funnel is the project?); and how compelling the investment case is. The idea is that we ascertain our best estimate of the expected value of future projects. The added benefit is that this value can be monitored and adjusted over time as the portfolio pipeline changes, to reflect the extent to which projects in the pipeline are converted to active projects and the ratio of forecast benefits to benefits realized."

"That kind of sounds like something we might find of value—if only I had any confidence I'll remember it all tomorrow!" complained John.

"Don't worry, I'll send you a summary—although please don't give it to those real options consultants. We have our reputation to protect, you know!" joked Bill.

"Fair enough. Where does that leave us?"

"With one final category of project—the mandatory ones that argue for investment on the basis of a legal or regulatory requirement or because business continuity depends on it."

"But surely, Bill, you don't expect people to manage the benefits from 'must-do' projects—the benefit is clear, we don't end up with failed systems or in jail!"

"Well, ignoring them is one approach, John, but it's not the one that we recommend. You see, the problem we often find with 'must-do' projects is that they are not the cheapest or indeed the best way of achieving the desired outcome, and often they are the 'Trojan horse' for a whole load of functionality that would not have passed the investment test otherwise."

"You're right!" exclaimed John. "I remember all that Y2K stuff; it was a field day for getting loads of IT replaced that actually had nothing to do with Y2K. But don't quote me on it; I'm 'poacher turned gamekeeper,' remember!"

"Okay, fine—just pay the bill on time! Seriously, what we recommend is an approach that makes the implicit value of such projects explicit. This is based on three pieces of analysis:

- First, a cause-and-effect analysis to demonstrate the rationale for linking the project to the business requirement, approved by the business sponsor, to provide a degree of confidence that the project will address the issue at hand effectively.
- Second, the key success criteria both in terms of final outcomes and 'leading' outputs, that will be used to measure success. For example, if we're investing to reduce risk, how will we know that risk has really been effectively mitigated?
- Third, an options analysis demonstrating that the net cost required represents the most cost-effective way of addressing the issue also approved by the business sponsor and the finance director.

Using these analyses, the governance bodies can come to an informed decision as to whether it is worth investing the requested funds to achieve the demonstrated impact."

"Okay, I think I've got that," said John. "I presume that takes us on to the third key?"

"You're right," Bill confirmed, "but I'd like to invite Chris in on this since she's done some work in this area recently and has made some real progress."

John put Bill on hold and called to Chris's phone. She picked up on the second ring. "Chris, can you join us for a few minutes? John could do with the benefit of your experience. I'll conference you into the call."

"Sure, John. Don't forget though, it's Friday afternoon and you still owe me from the last time," chided Chris.

"Okay, okay. Just put it on my tab," retorted Bill. "I'll conference you into the call."

Bill deftly pressed the right buttons and presto—they were all on the call together.

KEY 3: MOVING BEYOND BENEFITS REALIZATION TO VALUE CREATION

"Hi, John, how can I help?" asked Chris.

"I really appreciate this, Chris. I've been talking to Bill about the benefits paradox—we all invest in projects to realize benefits and yet we all seem to struggle to demonstrate that those benefits have been realized. Bill's been explaining to me that the answer lies in ensuring that a focus on benefits underlies our portfolio management processes with the objectives of ensuring that all benefits claimed are robust and realizable, and that all potential value is captured—I've got that right haven't I, Bill?"

"That's a pretty good summary, John."

"Okay, good, but Bill suggested there was a third key and that you're the expert to advise me, Chris."

"I'll see if I can do justice to Bill's confidence in me," joked Chris.

"Keys one and two have laid the basis for a value-driven approach to Project Portfolio Management. Ensuring we realize the full value from our portfolio, however, requires that we go further in shifting the focus from reactive benefits tracking to proactive value management. This in turn requires:

- "First, effective leadership that embeds the expectation that value will be actively managed in a constant search for ways to leverage investments. Portfolio management is a change program—senior management needs to 'walk the talk,' if you'll pardon the cliché.

- "Second, we need to change the organizational incentives. This includes realignment of the reward and recognition policies, to put people's focus on value creation as much as it is on project delivery.
- "Third, real accountability. Here we hold people accountable for the commitments they've made, but also with a focus on organizational learning. The objective is not to punish people for good, well-intentioned work that resulted in failure; rather, it's to ensure that lessons are learned. The emphasis should be on the future, not the past.
- "Finally, that the organization manages benefits from an enterprise rather than a project basis. This means that each year there is a clear plan about what benefits will be realized from the organization's portfolio—and this enables management to ask whether the fore-casted benefits are sufficient given the accumulated investment. No more cases of forgetting the investment once the project is completed, and this also helps achieve synergies across the portfolio—all those hours saved can be tracked and realized. Your portfolio management office can play a key role here, and I've seen some organizations expand the role to encompass that of a value management office charged with capturing emergent benefits and disseminating learning and best practice."

CONCLUSION

"So, let's summarize where we've got to," John recapped. "We need to focus on three things: first, ensuring all benefits claimed are robust and realizable; second, that we capture all forms of value created; and third, that we create value by managing benefits from an enterprise rather than project perspective. . . . Have I got it right?

"That's about it," replied Bill, "although it's easier to say than to achieve in practice. I reckon that napkin on which you wrote your notes at our first meeting must be pretty full by now—so let me email you something that Chris and I carry around with us. It captures the salient points on one page, just as your business cases and benefits reports should. I'm sending it now."

John looked at his in-box and opened the latest message and an attachment entitled "The Ten Principles of Effective Benefits Realization Management." "Got it, just give me a minute to read it . . .

The Ten Principles of Effective Benefits Realization Management

- Benefits must be placed at the center of the portfolio management and investment appraisal processes—funding should be linked to benefits forecasts and key stakeholders should be clear about what benefits they are buying.
- Benefits realization starts with the business case—ensure that the business case includes all activities and costs required to realize the forecast benefits.
- Funding allocations should be incremental and continued funding should be directly linked to the latest benefits forecast—regular checkpoints (stage gates and portfolio level reviews) should be built in so that if benefits fall away, budgets can be adjusted accordingly.
- Where possible, "book" the benefits early—by cutting budgets, headcount limits, and target unit costs, and by including them in divisional and individual performance targets.
- Optimism bias is a reality—benefits tend to be overstated and are often little more than unsubstantiated assumptions. Such claims must be robustly scrutinized and challenged.
- Benefits should be validated wherever possible to ensure they are realizable—by agreeing them with the recipients and those who will be responsible for delivering the business changes on which benefits realization is dependent.
- Capture all forms of value-added—efficiency (both time and financial savings), effectiveness (improved performance), foundation/ potential opportunity value, and the value represented by the avoidance of "things gone wrong."
- Benefits need to be actively managed—to ensure that forecast benefits are realized (especially important where those benefits are dependent on business change) and to capture benefits that were not anticipated at the business case stage.

"This is fantastic," exclaimed John. "I reckon this is just what I need to show Hannah because it takes us beyond process to governance and behavior."

"That's the lesson for the day," Bill replied, "unless you get the governance and behavior sorted, all the process stuff is just more bureaucracy.

If you remember at our first dinner, I emphasized that portfolio management is more than a set of processes; it represents a shift in the culture of the organization."

"I remember—and it's something that's become only too clear to me over the last twelve months." John confirmed.

Bill added, "By golly, I think you've got it, John!"

"Hey, I reckon that's enough of the humor at the expense of the rookie practitioner—I might be a slow learner but I think this is going to pay real dividends, real soon, and dinner will be on me again. So thanks, but I've really got to go. Hannah's car's still in the parking lot and I don't want to miss her."

"I'll just email you a couple more documents that you can read later," said Bill. "The first is the EPMC Working Document on Benefits Realization and it really sums up everything I've been talking about for you, John. The second one by the EPMC is a bit longer and goes into detail on operating considerations. It should answer a lot of questions you still might have. And I'll take you up on that dinner. Good luck talking with Hannah."

EPMC WORKING DOCUMENT ON BENEFITS REALIZATION

We invest in projects and programs to realize benefits in terms of increased revenue or sales; cost and time efficiency savings; adherence to regulatory and legal requirements; maintenance of business as usual; contributions to strategic priorities; and to achieve business performance improvements. But:

- In many cases these benefits don't just materialize automatically— realization is dependent on business change including staff training, business processes reengineering, and redeployment of resources.
- Organizations struggle to demonstrate achievement of the anticipated benefits and a positive return on their investment.

This is important because it undermines our portfolio prioritization processes (which rely upon accurate and reliable data) and means that we fail to optimize the return on our investment of shareholders' and taxpayers' funds.

Addressing this benefits puzzle requires that we use the Three Keys to Benefits Realization.

Key 1: Ensuring All Benefits Claimed Are Robust and Realizable

This is done so that the organization's portfolio management process has reliable data on which to select and prioritize potential investments, and to increase the probability that these benefits will be realized in practice. This requires that we:

- Establish a benefits framework—the set of rules about how benefits should be classified, quantified and valued, to provide a consistent approach for the preparation of investment cases across the portfolio.
- Validate benefits—wherever possible, and in particular, with the recipients.
- Embed a benefits focus in our regular project stage gate and portfolio level reviews—including regularly asking the question "Is the investment logic still valid?" and gaining formal recommitment to the benefits case by those who will be responsible for realizing them.

Above all, be clear about the benefits you are buying and the measures that will be used to assess realization, and bring all this together in a Benefits Realization Plan for each project.

Key 2: Capturing All Forms of Value Created

So that projects don't just stop at the hurdle rate, recognize the full potential benefits available from an investment. This in turn requires that we identify:

- Efficiency benefits, both time and financial savings—and book them in budgets wherever possible, while remembering that:
  - There need to be checks that the forecast benefits have been realized rather than just top-slicing budgets so impacting negatively on output and service quality.
  - Time savings are a *potential* benefit—the value is only realized when the time saved is used for some value-adding activity.
- Effectiveness benefits, in terms of improved performance and contribution to strategic priorities. Using Strategic Contribution Analysis, combining strategy mapping (from vision through strategy to success measures) with benefits mapping, so that we are clear about

the link between project benefits and the success measures of business strategy.

- The flexibility or options value that infrastructure investments provide in the ability to exploit applications in the future that are dependent on that infrastructure—based on a "Potential Opportunity Value" approach in which a proportion of the value of the portfolio pipeline is taken into account based on an assessment of the probability that the applications that will deliver these benefits will come to fruition. Performance should be monitored in terms of:
  - The extent to which projects in the pipeline are converted to active projects
  - The ratio of forecast benefits to benefits realized
- The value of legal/regulatory compliance and maintenance of business as usual—this value can be assessed by making the implicit value of such investments explicit by presenting the portfolio governance bodies with three pieces of analysis:
  - A cause-and-effect analysis demonstrating how the project meets the requirement
  - The key success criteria that will be used to measure success
  - An options analysis demonstrating that the net cost required represents the most cost-effective way of addressing the requirement

Using these analyses, the governance bodies can come to an informed decision as to whether it is worth investing the requested funds to achieve the demonstrated impact.

Key 3: Realizing Benefits and Creating Value

This lets us realize the full value from our portfolio by shifting the focus from reactive benefits tracking to proactive value management. This in turn requires:

- We manage benefits from an enterprise rather than a project basis. This facilitates a continued focus on benefits after project closure and achieving synergies across the portfolio.
- Leadership—embedding the expectation that value will be actively managed in a constant search for ways to leverage investments.

- Changing the organizational incentives, including realignment of the reward and recognition policies with value creation as well as project delivery.
- Accountability with learning—the emphasis should be on the future, not the past.
- Expanding the portfolio management office role to include the functions of a value management office where equal focus is given to the five fundamental questions of enterprise portfolio management:
  1. Are we investing in the right things?
  2. Are we optimizing our capacity?
  3. How well are we executing?
  4. Can we absorb all the changes?
  5. Are we realizing the promised benefits?

Part III

OPERATING CONSIDERATIONS

Chapter 8

The PPM Process

To better understand Project Portfolio Management, let's look at some of the key aspects of the PPM process. First, where does PPM fit with other business processes in a company? Figure 8.1 graphically depicts how PPM relates to a variety of business processes in an organization. The PPM process is located in the center of the graphic only in order to highlight its various key relationships to other business processes. (PPM is not the center of the business universe.)

Foundational Principle

PPM works with other business processes and disciplines—it doesn't replace them.

The PPM process provides the forum, discipline, decision criteria, funding, and decision-making authority to effectively manage the portfolio of projects and programs.

Foundational Principle

PPM provides the forum, discipline, and decision criteria for effectively managing a portfolio of projects.

PPM Process Context

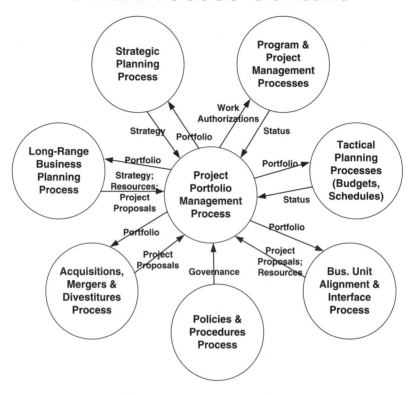

Figure 8.1   Project Portfolio Management is one of many interrelated processes that make up the complex unit that we call a business.

PPM COMPONENTS

Let's look briefly at some of the major components and deliverables associated with PPM. To do this effectively, let's consider a high-level process flow (see Figure 8.2). Our example PPM process comprises twelve steps:

1. Create project proposal
2. Gate 1: Approve project proposal
3. Incorporate into budgeting process
4. End/suspend or replan project proposal
5. Develop project management plan and business case
6. Gate 2: Authorize implementation
7. Analyze portfolio and recommend project priorities
8. Prioritize project portfolio

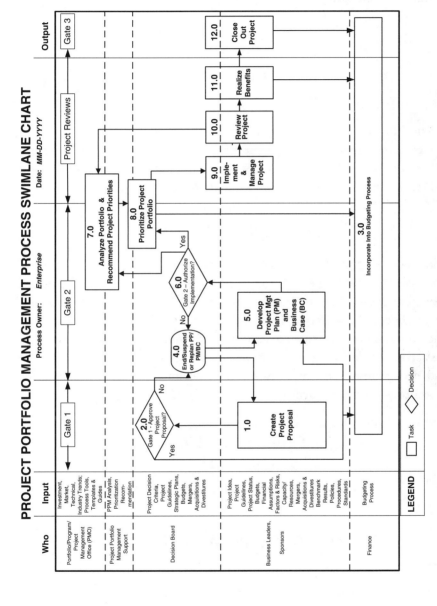

Figure 8.2   This twelve-step process flowchart highlights the major deliverables, decision points, and participants in the PPM process.

9. Implement project
10. Review project
11. Realize benefits
12. Close out project

Foundational Tool

PPM process flowchart

The PPM process itself is usually managed by the PMO (the project, program, or portfolio management office). Inputs from the PMO throughout the process include information on investment, market, technical, and industry trends, as well as process tools, templates, and guides.

THE PROJECT PROPOSAL

The process begins with an idea documented in a project proposal (Step 1). The project proposal usually comes from business leaders and/or project sponsors. Inputs to the proposal include project ideas, guidelines, and status; budget information; financial assumptions; risks and other factors affecting the project; resource capacity information; benchmarking information on similar projects; information on relevant mergers, acquisitions, and divestitures; and information on relevant organizational policies, procedures, and standards.

Foundational Tool

Project proposal

PROJECT PROPOSAL APPROVAL

The project proposal then undergoes scrutiny in Gate 1 of the process, where the project proposal is approved or rejected (Step 2). (Note: approving the project proposal is not the same as approving the project, which doesn't happen until Step 6.) This approval process is undertaken by a decision board. Inputs to the approval process include project decision criteria, project guidelines, strategic plans, and budget information.

We recommend the use of standard presentation templates for each gate in the process, which produces the best opportunity for the decision board members to view each deliverable on equal footing. It also prevents great "sales jobs" from being presented as proposals or business cases.

The decision board members view the proposal through the filter of the decision criteria they maintain. The decision criteria might include such items as the availability of people required to conduct the project or the financial costs and benefits (at a rough-order-of-magnitude level). Other possible criteria include the strategic alignment and the risk, or probability of success, of the project (see Figure 8.3). The project proposal

Project Portfolio Management
Prioritization Criteria
for Project Proposals & Business Cases

Criteria	Weighting (1 to 10, L to R)
Value	
1. Cost Savings	
2. Cost Avoidance	
3. Return Rate for <xx> Months	
Strategy	
4. Enterprise	
5. Business Unit	
6. Organization	
Balance	
7. Intangible (non-dollar) Benefits	
8. Probability of Success	
9. Resource Availability	
Linkage	

Figure 8.3   Various decision criteria are used to evaluate the value of a project and in determining whether to add the project to the organization's active portfolio.

approval process also includes determining the project's dependency on, and relation to, other projects in the portfolio.

If the project proposal is not approved, the proposal can be withdrawn, put on hold to revisit at a later date, or sent back to the proposal team for replanning and resubmittal (Step 4).

THE BUSINESS CASE AND PROJECT MANAGEMENT PLAN

If the project proposal successfully navigates through Gate 1, then the decision board sends the proposed project information to the finance department (Step 3) so they are aware that this project may need to be added to the budget if authorized in Step 6. The decision board also allocates funding for the development of a formal business case and project management plan (Step 4).

Foundational Tool

Business case

The project management plan usually consists of several deliverables. Which deliverables, as well as the amount of content in them, depends on a variety of factors, including project size, complexity, scope, cost, and benefits expected. Possible deliverables include:

- Statement of work
- Project charter
- Work breakdown structure
- Schedule
- Responsibility assignment matrix
- Communication plan
- Budget plan
- Risk management plan
- Change management plan.

Foundational Tool

Project management plan

PROJECT PRIORITIZATION

If a proposal makes it past Gate 2, then it goes into the project portfolio "hopper" or holding area. The decision board reviews the projects in the hopper together with those already part of the active portfolio. The board takes a holistic view of the portfolio to see how the projects fit together and how well the portfolio is balanced in order to meet the organization's strategic goals.

The projects in the portfolio are then prioritized according to the scores they receive based on criteria maintained and weighted by the decision board. The scoring criteria can be simple or complex, and may be the same as the decision criteria used in determining whether to fund the project. A score is given for how well the project meets the criterion, say on a scale from 1 (not well) to 10 (extremely well). The criteria may also be weighted based on how important a particular criterion is to the organization. Criteria may include such factors as fit with mission, consistency with objectives, consistency with strategy, contribution to goals, risk level acceptability, and so on. The scores are totaled for each project in the portfolio and put into priority order, the highest-scoring project first.

These scores equate to the "facts" as we know them (or forecast them). The facts determine a priority order based on the expected contribution of each project to the overall portfolio, but people still make the final decisions about the actual priority order of the projects in the portfolio. So, with facts in hand, the members of the decision board use their guts to determine the final priority list. Sometimes they're in agreement, but sometimes they aren't. And that is when the fun really begins—and the decision board really has to work hard to come to a collective agreement on the portfolio.

Foundational Principle

Portfolio prioritization scoring models provide facts and data, not decisions. People make the final priority decisions. The models enable informed decisions.

To assist the decision board in doing their work effectively, we recommend a simple, conceptual tool, the portfolio equalizer.

Foundational Tool

The portfolio equalizer

The portfolio equalizer metaphorically brings the technology of sound into the realm of PPM. Just as a stereo's equalizer balances the various aspects of the sound waves to produce the desired fidelity for optimal performance, so does the portfolio equalizer. The knobs and slide switches on the portfolio equalizer produce the portfolio balance required by the business strategy, market situations, and emergent conditions. The knobs and slide switches can be adjusted by the decision board as needed. The result is a possible reordering of the project priorities based on the new application of weights to the facts followed by one last comparison of facts-to-guts and a decision by the board on the final prioritized list of projects. This conceptual tool has proven effective in our implementations and operation of PPM in multiple businesses.

PROJECT AUTHORIZATION

Once the proposal team develops the business case and project management plan, they return to the decision board (Gate 2) for formal authorization (Step 6). If the project proposal (with accompanying business case and project management plan) is not authorized, the proposal returns to Step 4, where, again, the proposal can be withdrawn, put on hold to revisit at a later date, or sent back to the proposal team for replanning and resubmittal.

PROJECT EXECUTION AND REVIEW

If the project is formally authorized, the next step is for the project portfolio support folks to analyze the project portfolio and develop a recommendation of priorities (Step 7) based on scores and project dependencies and constraints for consideration by the design review board (DRB). The DRB will examine the facts as we know them, the recommended priorities, and the project dependencies and constraints, and then work toward an agreement on the priorities of each project in the portfolio. This is where the value of discussion and consensus comes into play. The scores themselves do not always dictate the actual priority of each project. Rather, the scores are indications of how the projects relate to one another based on the importance, or weighting, given to the criteria used to evaluate the projects and portfolio as a whole. Once agreement is reached by the DRB, then the priorities of each project are recorded and published (Step 8).

After prioritization is determined, then the finance department is notified so the project is entered into the organization's budget, and the project is ready for execution (Step 9). Throughout its execution, the project's performance is monitored (Step 10). The project often must pass through decision-gate evaluations to determine whether to continue with the project, to put it on hold, or to kill it altogether.

Project reviews involve a reverification of the project's critical success factors—including resource availability and the continued validity of the business case—with the business sponsors. In addition, shifting business, technology, and market conditions can rearrange priorities. The decisions made in response to these shifting conditions often result in a realignment of the project portfolio, which may or may not affect other projects in the portfolio. Replanning may be required, including changes in resource allocation and scheduling.

As the project comes to completion, a process is put in place to document whether the benefits documented in the business case are actually realized (Step 11). This process may take a significant amount of time to complete based on the type of project and the timeframe for which benefits may actually occur. But the process must be put in place to actually document this information before the project comes to a close. This information on the individual project is added to the documentation used to analyze the performance of the entire portfolio.

Finally the project is closed (Step 12). This process insures that all of the other processes required have been completed and the completed product is handed off to others, or simply closed if the project has been cancelled. Again, this information on the individual project is added to the documentation used to analyze the performance of the entire portfolio.

CHAPTER SUMMARY

Foundational Principles

- PPM works with other business processes and disciplines—it doesn't replace them.

- PPM provides the forum, discipline, and decision criteria for effectively managing a portfolio of projects.
- Portfolio prioritization scoring models provide facts and data, not decisions. People make the final priority decisions. The models enable informed decisions.

Foundational Tools

- PPM Process flowchart
- Project proposal
- Business case
- Project plan
- The portfolio equalizer

Chapter 9

Setting the Foundation for Success

A successful implementation of project portfolio management is a journey and not a destination, a marathon and not a sprint. To that end, one of the immediate needs to begin the journey is a case for change, or, in more formal terms, a business case for PPM, with sponsorship, alignment, and ultimately approval. Of course, there's usually recognition that there is a reason or need for PPM that comes about because of inefficiency, ineffectiveness, or a general lack of business value from the existing investment in people and financials/budget. These are the most common reasons for change and can be related to by all levels of the organization, from the highest to the lowest. Identifying, recognizing, and defining these reasons for change are the foundation of the business case.

Foundational Principle

The successful implementation of PPM is a journey, not a destination.

THE BUSINESS CASE FOUNDATION

There are many reasons why the business case becomes key to success. One primary reason is to gain alignment from all levels of the organization and not just have the case for change come authoritatively from the top of the organization. Associates at all levels of the organization should agree that the implementation of PPM will offer some beneficial results—for each functional area, for individuals, and, of course, for the organization as a whole. The answer to the question "What's in it for me?" becomes important so there can be a general agreement and alignment around the reasons for implementing PPM. This general agreement about the reasons for implementing PPM will help in promoting the engagement model and becomes a key enabler for the success of the organizational change management that will need to occur. The business case should also tie back to the organization's strategy so that there is alignment on the approach to successfully achieve the business results defined by the strategy. If employed and cultivated properly, PPM can become a key enabler for the organization in achieving its strategic goals.

Foundational Principle

The business case for PPM should tie back to the organization's strategy.

As with any project, the PPM business case should include data and information to answer the following:

- Why are the funds needed (including sufficient information to help weigh this project against other needs)?
- What is the business value of the proposed investment, both tangible (financial) and intangible?
- How does this project align with the organization's strategy?
- What are the root causes of the situation (include an analysis of the current situation and an assessment of options)?
- Who are the stakeholders for this project proposal?
- What are the relevant environmental considerations or factors?
- What are the risks of doing the project (and not doing the project)?
- What are the contingencies and dependencies?
- What are the specific recommendations for action?
- How will we measure success?

THE BENEFITS OF PPM

The potential benefits of pulling together a successful business case and, ultimately, successfully implementing PPM are:

- More of the "right" projects and programs being undertaken in terms of greater financial benefits and contribution to strategic targets and business priorities.
- Fewer of the "wrong" projects (as well as duplicate ones) started or more killed off earlier via active management of the project portfolio.
- More effective implementation of projects and programs via management of the project pipeline, project dependencies, interdependencies, and constraints.
- More efficient utilization of resources (people and assets)
- Greater benefits realization via active approaches to exploitation of the capacity and capability.
- Improved accountability and ability to demonstrate adherence to corporate governance requirements.

Foundational Principle

The successful business case for PPM documents the benefits it can bring to the organization.

These benefits are backed by research evidence as well:

Kendall and Rollins (2003) identify four generic problems associated with the absence of Project Portfolio Management: too many projects; projects that do not add value; projects that aren't linked to strategy; and an unbalanced portfolio.

Cooper (2005) concludes that "exceptional performance in product development is no accident. Rather it is the result of a disciplined, systematic approach based on best practices." Organizations that adopt these practices were found consistently to outperform the rest in the new product development arena. Cooper asserts that "numerous studies have confirmed that there is no direct link between a company's increase in spending in R&D and their success rate with new products. What then, if not spending, drives new product success? Significant productivity gains (in NPD) are possible through astute project selection. In fact, top performing businesses are four times more likely to deploy such practices, namely effective portfolio management."

Researchers from London and Ashridge Business Schools (Reyck et al., 2005) undertook a study that sought to assess the correlation between the application of Project Portfolio Management processes and techniques, and improvements in project performance. The study was based on a global survey of 125 medium-to-large companies. They found a strong correlation between increasing adoption of Project Portfolio Management processes and a reduction in project-related problems as well as project performance. That is, PPM is correlated with improved project performance, and this link increases with the maturity of the processes adopted: *As organizations increasingly adopt PPM approaches, the impact is strengthened.*

Lesson Learned

As organizations improve their PPM processes, project performance improves.

Research from Gartner (McClure, 2007) demonstrates the impact that successful implementation of IT PPM can have. For example, "there is a definite value payback associated with well-run portfolio management practices. It is not uncommon for IT portfolio reviews to identify cost savings in the total IT budget potential ranging from 10% to 20%. These initial savings amounts often stem from application and software licensing sharing opportunities, common application sharing and infrastructure consolidations, and stopping or delaying additional funding for poorly performing projects. More importantly, overall risk management can help to ensure better performance impact on desired agency program or business outcomes." Matt Light and colleagues (2005) argue, "Many IT organizations do not have the skills or the discipline to avoid being swamped by the flood of proposals and projects. Some enterprises adopt PPM mainly driven by the need to prioritize; others want to halt low value projects, or those that falter or overrun budgets and schedules, or whose business case lapses. When fewer projects are prioritized, and troubled ones are terminated, the freed up resources can then be reallocated to higher-value initiatives. In other cases, early visibility into troubled projects enables intervention and remediation to avoid major disruptions to the business. This gives enterprises a choice between obtaining the same value for less investment or more value from the same investment." Di Maio (2006) concludes that their research demonstrated the benefits of adopting an IT portfolio management process in terms of improved alignment with political and strategic priorities. The absence of PPM was seen as "the most fundamental impediment to aligning IT activities with business and political priorities."

According to Butler Group (September 2005), the benefits of PPM include improved communication between IT and the business and involving senior management in the project selection process; providing transparency about what is being spent, where, and why; addressing the issue of too many projects, weak ones, and nonstrategically aligned ones, thus maximizing value; and identifying risks. Butler reports that Accenture has found that a structured process to investment management can enable savings of 10 to 15 percent in the IT budget within one year and better decision making can improve IT productivity by up to a further 20 percent.

So, one-off savings on the order of 20 to 30 percent of total project spending are common *when PPM is implemented* (i.e., from removal of low value, duplicate, redundant, and poorly performing projects). This is relatively easy to measure; what is not so easy to measure is the impact of PPM going forward when many of the benefits will be nonfinancial (although they may well ultimately contribute to improved financial outcomes) or difficult to measure (what is the value of poor projects that are not started, for example?) or difficult to attribute to PPM alone (rather than improved project management). Consequently, the emphasis may well be on proxy measures such as management satisfaction and process compliance.

At the end of the day, we want the business case to be approved through the compelling argument (and need) that is defined with a promise of a measureable result for the investment—the return on the investment. Benefits realization of the business case should be tracked, with relevant baselines and measures, as the implementation is delivered and matures over time.

With the business case approved and funding secured, the next phase of foundation setting is ready to be initiated. In our experience, the foundation for a successful implementation of PPM can be traced back to the balance within the triad of people—process—technology. This balance depends on many factors within the organization as well as the environment in which the organization lives. Some of these factors will be discussed later in this chapter.

Foundational Principle

PPM success requires a balance of people,
process, and technology.

THE PEOPLE FOUNDATION

You can have great processes, you can have great technology but you still need people to be engaged and empowered for success. One of the most obvious places to start with the foundation is with the identification of a sponsor and the related stakeholders. No doubt, these are important to the foundation, but the people who will be impacted by the initiative—those who are the project managers, those who are the providers of information and data, those who are affected by the decisions made as a result of the data, process, and technology—are in many cases equally important, if not more so. As has been highlighted throughout the preceding chapters, implementing a PPM initiative should not be taken on lightly and will likely impact many aspects of the organization—how it functions and the people who perform within it. PPM is not as easy as many portray it to be.

Foundational Principle

Engaged and empowered people are keys
to PPM success.

The players that typically make up the people foundation are the decision board or senior leadership, the sponsor, the stakeholders, the project leader, the project team, and the people impacted by the initiative.

These are the typical people players for implementing PPM. The people players change or evolve when the implementation ends and the PPM function takes its place operationally. As PPM moves to an operational state, the players typically evolve to a PPM leader at the vice president or director level, the portfolio manager(s), the portfolio analyst(s), and the people impacted by the initiative.

Ideally the sponsor should be someone with some degree of authoritative or position power to influence and ensure an environment for success exists. The authoritative or position power concept is important because by nature people are resistant to change. It's been proven time after time that there needs to be an impetus to change, and many times the change has to be pushed by some degree of positional power. The sponsor needs to be careful how they use their power for change, employing it at appropriate times with the most impact. The sponsor also needs to play other important roles as well with their teams and other stakeholders.

Foundational Principle

The PPM sponsor should have enough authority to ensure an organizational environment exists to enable success.

In more hierarchical structures the stakeholders should have breadth across the organization. That is, the functional areas of the organization should be represented in some way so that their differing perspectives can be heard and accounted for in the decision-making process. Stakeholders should work with the project team to ensure that a holistic approach to the initiative is taken so that each of the functional areas of the organization and the business are accounted for and represented. The sponsor and the stakeholders need to be educated on PPM—what it is, what it does, and why it's needed for the enterprise. This education is critical so that they have a clear understanding and speak intelligently of the benefits and changes needed to support the PPM initiative in their critical roles. The education should occur before the initiative kicks off.

The project leader and project team that is tasked with the implementation of the PPM initiative are positioned directly in the critical path. The team works with the sponsor and stakeholders on one end of the spectrum as well as the people impacted by the initiative on the opposite end of the spectrum. The project team is expected to run the initiative as a program/project with the typical responsibilities of managing the project plan, the budget, the process definition, and the technology implementation. These are givens. However, the people perspective— more formally defined as people or organizational change management— is often overlooked or deemed less important than the more familiar project management tasks. To address this foundationally important piece of the implementation, the project team needs to be aware of and

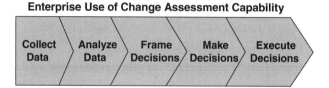

Enterprise Use of Change Assessment Capability

Figure 9.1 The PPM project team can assess the organization's capability for change and use that information to aid the change management necessary to effectively implement PPM throughout the organization.

plan for people change management by engaging both ends of the people spectrum. Some implementations should consider the opportunity to engage a professional consultant who is trained and experienced in the art and science of organizational change management. The project team can perform some of the fundamental tasks of change management, including assessing the organization's experience with change, openness to change, and ability to change—change impact versus change capacity.

The sponsor, stakeholders, and project team need to have an honest assessment of the organization and its environment. The result of this assessment should determine whether a change management consultant is needed and/or to what degree they need to be employed. Ideally the consultant should be a part of the team from the inception of the project as well as a part of the assessment of the organization. This early engagement will allow for an appropriate level of support needed to implement PPM and, more importantly, make it sustainable and viable for years to come.

Foundational Tool

Organizational assessment.

"Early engagement" emerges as a key organizational change management concept in the implementation and eventual success of PPM in the organization. Stakeholders need to be engaged early and invested in the process through their responsibility to their functional area. The focus of engagement needs to be on the people who will be impacted by the initiative—the project managers, the providers of information and data, and those affected by the decisions made as a result of the data, process, and technology.

Lesson Learned

Stakeholders need to be engaged early and invested in the PPM implementation process.

Use an organization map to identify key thought leaders, influencers, early adopters, as well as potential resistors in the organization. Interviews of these key members of the organization should be conducted on an individual basis and by functional area. Initially the communication of the PPM initiative should focus on educating the group on what it is and why it's important to the organization. During the interviews, establish a dialog. The dialog is critical so that, as the project progresses, the feedback loop will help the project team understand where things are going well and where attention needs to be applied as well as who is an adopter/supporter and who is a resistor. It's also important to identify the WIIFMs (what's in it for me) for all the stakeholders. That is, what they gain by engaging in and adopting the PPM processes and technology. Of course, the organization's environment and culture will impact this engagement because some organizations choose to implement PPM with a top-down approach driven by senior management. The effect of early engagement also carries over when the PPM initiative transitions to an operational state. The dialog/feedback loop needs to be nurtured through the transition so enhancements to the PPM processes and technology in the future can be supported as well.

Foundational Tool

Organization map

To implement PPM successfully, usually a PPM implementation team needs to be created to deliver the new capability. We'll discuss the team in more detail in the subsequent process and technology foundation sections, but we recommend an integrated team approach with three sub-teams — one each for People, Process, and Technology. An overall implementation leader should be appointed to oversee and coordinate the implementation and sub-teams for each of the aforementioned areas.

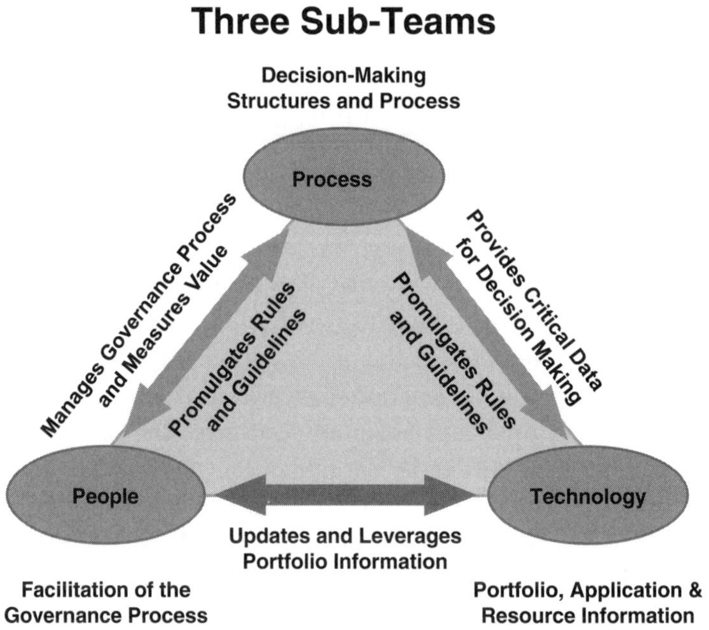

Three Sub-Teams

**Decision-Making
Structures and Process**

Process

Manages Governance Process and Measures Value

Promulgates Rules and Guidelines

Provides Critical Data for Decision Making

Promulgates Rules and Guidelines

People   Technology

**Updates and Leverages
Portfolio Information**

**Facilitation of the
Governance Process**

**Portfolio, Application &
Resource Information**

Figure 9.2   The PPM implementation team comprises three integrated subteams—people, process, and technology teams—to deliver PPM, as well as the required organizational changes, successfully.

Work done in the implementation phase of the initiative determines the path to successful organizational change management because the implementation approach relies on the engagement of the stakeholders in defining the desired future state processes and technology.

The people team needs to assess the organization's readiness/willingness to change—its change capacity. Based on the assessment, a plan should be developed to address the functional areas, or even individuals within the organization, that will struggle with or not be open to change. The plan must cover the communications and training needed to address the impact of the change. Once the initial communications have been kicked-off and the process and technology teams begin implementation, the people team will turn its focus to communicating the progress of the implementation to stakeholders, possibly identifying additional needed subject matter experts or key thought leaders—all in keeping with the theme of organization engagement. There may be change in control and decision-making power that will need to be identified and dealt with or even escalated to the stakeholders and sponsor for resolution.

As the implementation workshops get under way, the people team will also need to develop training materials. The materials can be leveraged directly from the outputs of the process and technology teams. How others in the organization are going to be trained is another consideration—will they require classroom training or computer-based training; will a new PPM tool be introduced; is there a need for training aids? Planning for beyond the implementation of the PPM solution should be considered as well. The materials developed to support the initial rollout should be maintainable for future upgrades and enhancements, but they should also be made available as reference materials for any member of the organization who needs to be refreshed on how to perform a task or what the expectations for roles and responsibility are.

A lesson learned—create a knowledge base that includes this type of training and reference material and enhance it with policies and procedures so there is a standard repository. A collaborative environment for using the knowledge base can be accomplished through an Intranet portal, which could serve as a key communications vehicle with the program/project management office (PMO).

Lesson Learned

Create a standard repository of PPM training materials, including policies and procedures.

Additional information about the process and technology foundations follow in the subsequent sections as well as the continued discussion as to how the three teams work in concert to deliver the initial implementation of Project Portfolio Management and a strong foundation for ongoing success.

THE PROCESS FOUNDATION

It is important to have a good understanding of what the existing process is and how the process was put into place. When trying to set up the foundation for process success, it's best to learn as much as possible from what has worked previously and what has not. There are lessons in those experiences that should afford any new initiatives with a significant head start in avoiding processes that did not work well previously. Even though the organization's environment may change and people in the organization may change, the lessons learned will still be valuable. To that end, it would even be advantageous to tap into people in the organization that have some history with the process.

Assuming that the key success factors described in the previous chapter have been performed—the sponsor team has conducted a kickoff meeting and a team has been identified to work on the PPM project—it's time to address the process foundation. We recommend using a process-based approach to help establish a solid foundation for process success. A three-step process model—defining the current state, identifying the issues and opportunities, and then describing the desired future state—is an effective means of establishing the process foundation. In the first step, where the current-state processes are identified, the primary objective is to ensure that all of the current processes and their intended benefits and reasons for existence are discussed and understood. At first glance this would seem to be a bit pointless. However, in practice it's usually an enlightening experience. An effective means of completing this task is to conduct a series of workshops with the team and subject matter experts.

Foundational Principle

Establish the process foundation in three steps: define the current state, identify issues and opportunities, and describe the future state.

Several key activities should happen prior to the start of these workshops. One is deciding on the composition of the team that will participate in the workshops. The PPM team leader plays the important role of facilitator for the workshops and is supported by a team. The facilitator is responsible for keeping the team on schedule based on the project plan, which usually means the facilitator will have to encourage discussions at times and cut some off at times. The facilitator will need to keep members of the team engaged and not let any one person or functional area dominate the development of the process. This is clearly not an easy task; it is important to appoint a facilitator who has experience, and it may even be advantageous to use a consultant in the role.

Now the team. The process team should consist of representatives of each of the major functions within the organization (program/project management office, finance, sales and marketing, operations, and so on). The representatives from these major functions should represent their perspectives, their roles, and their needs so that they can be understood and integrated into the future-state process. The team structure also plays into the key tenets of organizational change management that is so important for the success of the PPM initiative. By having each of the major functions represented and engaged, they all feel ownership and attachment to the future-state process that is ultimately defined. Other potential candidates for the team include stakeholders, process owners, and even resistors. The resistor is one who is not easily moved to change and in general is known to want not to change—every organization has them. If engaged properly and nourished, the resistor can be turned into a huge advantage as an advocate.

Now that the team is identified and formed, but before delivering the workshops, the attendees should be asked to gather documents containing important and relevant information such as:

- Process maps
- Process flows
- Roles and responsibilities
- Governance policies
- Procedures and controls

Once the material is pulled together, it's time to schedule workshop number one to review what information was collected. At the start of the first workshop, the team lead should work with the attendees to set expectations for the workshop, to establish roles for the attendees (facilitator, contributor, subject matter expert), and to gain agreement on team behavior (respect the opinions and feedback of the team, commit to no distractions). Once set, these principles should be used by the team lead to facilitate the workshops and keep the team focused on the tasks at hand.

Workshop number one can then transition into a presentation and discussion of the process instruments and materials collected. This could be a very short discussion if there is a low level of process maturity in the organization or quite lengthy if there is a higher level. Often there are

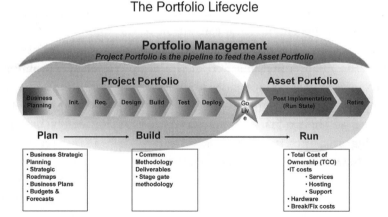

Figure 9.3   Visual communication devices like this graphic (and the following graphic) are useful in showing process flows and gaining team understanding in early process foundation workshops.

significant revelations during this first workshop because some attendees may not be aware of a process, the intended benefits of a process, or the effectiveness of a process. These are good revelations to have, however. They are good because it highlights a clear opportunity for the team to come together and create a truly integrated process—a key tenet to successful Project Portfolio Management. It is important that the workshop stress visual communication to keep the attendees engaged, especially with the process flows. We suggest making the process visible with large maps designating integration points with subprocesses or other critical path and dependent processes.

Another good communication technique is to print out the process maps on paper in a large format like wallpaper. This approach affords the team leader the opportunity to give the attendees pens and sticky notes to mark the process map identifying the areas where they see a need for improvement (red sticky note) or highlight an area where the process is performing well (green sticky note). Also, this gets people out of their chairs and engaged making them feel like they are truly part of the process. People are visual by nature and the picture of a process flow is helpful to focus the discussions.

In clearly identifying and understanding the existing process, the workshop discussions should highlight areas in need of improvement or areas where process does not exist (but is identified as needed). In each workshop the team leader should document these discussions with

Example of Future State Process Map

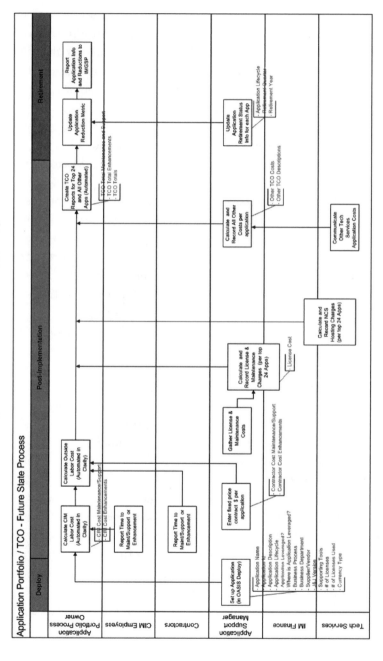

Figure 9.4 Print process maps in large format to help team members identify areas where they see a need for improvement or highlight an area where the process is performing well.

outcomes identified to form the basis for the future-state process. As the process possibilities are discussed, subteams can be formed to improve the process flow and/or add related subprocesses. And again we recommend using highly visual communication techniques in building the process maps. The subteams then present their work results to the team for discussion and this iterative process improvement continues until there are agreed upon and integrated processes. Other factors to consider during the process workshops include:

- Categorize processes as those that drive/deliver business value, those that drive efficiency and effectiveness, and those that are mandatory (regulatory).
- Commit to measures. Establish a baseline for whatever process needs to be measured and then establish the mechanisms (either automated or manual) that will provide the necessary information to produce the measures and indicators on a dashboard and/or scorecard. Strive for well-defined measures and key performance indicators (KPIs).
- Identify the drivers and the linkage with the organization strategy and the overall company strategy.
- Seek to understand where attention and rigor need to be applied versus where it will be unnecessary. Process needs to be an enabler and not a bottleneck.
- Consider applying Lean Six Sigma theory.
- Consider the use of some Six Sigma tools that can prove to be very helpful such as a FMEA (Failure Measures Effect Analysis), CTQ (Critical to Quality), or VOC (Voice of Customer).
- Establish a level of process governance commensurate with the need in the organization.
- Establish clear roles and responsibilities for the processes and how the organization will work with them—use a swim lane diagram for the roles as a part of the process maps.
- Use an integrated process for project identification, project classification, project evaluation and selection, project prioritization, and the full project lifecycle through retirement.
- Establish data governance for the completeness and accuracy of information and who is accountable for it.
- Develop linkages to the resource portfolio and application portfolio processes to be integrated with the project portfolio processes for an integrated PPM approach.

- Consider capturing and tracking benefit realization and the related processes needed to do it.
- Commit to continuous improvement and dialogue with stakeholders.

Foundational Tool

PPM process flowchart

As the workshops progress over several weeks, the team will move iteratively from the definition of the current state to the identification of issues and opportunities and then the desired future state. The evolution of the process model into the desired future state needs to be an integrated model that should tie macro-level processes together and highlight the dependencies on supporting subprocesses. It is important to remember that the processes need the participation and accountability of the people in the organization to be successful.

Foundational Principle

Use common, standard terminology in communicating about PPM.

One additional consideration that proves to be critically important is to strive for a consistent and well-communicated language of portfolio terms and process—the common language of the organization. This will help to ensure that the organization is trained consistently, able to interact, and to govern for accountability. Not speaking a "common language" of Project Portfolio Management in the organization can drive misalignment and misunderstanding among groups or functions within the organization.

The outcomes of the process foundation, as with any process, need to be monitored against any changes in the organization such as changes in strategy, people, environment, and so on. The commitment to continuous improvement through dialogue with stakeholders and the people who helped design the future-state processes can also help ensure that a forum for changes is also available.

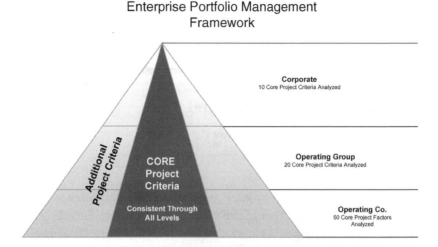

Figure 9.5   To facilitate understanding and consistent practice, organizations need to use common language in developing their PPM foundation.

As the process foundation continues to evolve out of the workshops, the people and technology foundations also evolve along with it, sometimes as a support and sometimes as a benefactor. This approach ties nicely back to the people foundation because the workshops are engaging people (stakeholders, process owners, resistors, and so on) in the process. At the very least, you will get a good sense for who is in alignment and who is not so that the organization change management plan can be targeted appropriately.

Again note the linkage and interaction between the three foundations as they continue to progress toward the delivery of the PPM framework.

THE TECHNOLOGY FOUNDATION

Building off of the previous foundations, the technology foundation seeks to enable process efficiency and to enhance the effectiveness of people in the pursuit of providing business value. The alignment of the objectives of the PPM initiative with the appropriate level of technology is an important initial step within the Technology Foundation. One of the very first decisions that should be made with regard to technology is how much is needed. It's not easy to answer this question, but some implications to be considered again relate back to people and process.

Foundational Principle

Alignment of PPM objectives and level of technology is critical.

With regard to process, the answer depends on how much is being proposed, how complicated it is, and how much effort is needed to integrate it within the organization. A high degree of process complexity and the need for integration between multiple processes (project, application, and resource) point to a need for a robust technology solution. A low degree of process complexity with a straightforward process approach may point to a simple technology solution, leveraging office automation software packages already installed on most computers. Determining the right amount of technology with the appropriate sophistication of tools is a choice that should be made as an investment with proper analysis and rigor.

With regard to people, the answer depends on the size of the organization and the scope of the PPM project. Is the PPM initiative a global implementation or a local one? Is there a big need for extensive data collection, analysis, and reporting? Is there an existing acumen in the organization to set up, run, or support a tool? Are the users of the tool savvy enough to be effective in using the tool? All are relevant questions in identifying how robust a tool is needed. You don't want to buy complex technology at huge expense and then need a significant amount of instruction when something simple is really all that's needed to effectively accomplish the task at hand.

Foundational Tool

PPM software.

There are numerous PPM tools in the marketplace today and we don't endorse any specific one. These tools have been rapidly evolving over the years as Project Portfolio Management continues to grow in importance and prominence in most organizations. There are several criteria to consider when evaluating and selecting a tool:

- Since our foundational approach accentuates the process, it is important to select a tool that is configurable—that is, it can be altered or easily changed to enable and automate process. Some tools are not flexible in this regard and only follow a predetermined process for automation, so be sure to evaluate this important aspect.
- A tool with workflow automation is effective and can help to enable governance of process and notifications. Automated workflow can be used in instances where stage gates are employed within an organization—a request can be made, the governors of the process can be notified, the request evaluated, and the reply sent, all within the system while maintaining an audit trail of the decisions.
- The user interface must be intuitive and easy to use—various people will often use the tool, entering as well as receiving information. If the tool isn't easy to use, it's a barrier to the input of information that can ripple into poor quality or incomplete data and thus poor decision making.
- The ability to integrate data for a complete and integrated portfolio management perspective—project portfolio, resource portfolio, and asset portfolio—is another important consideration because decisions are often made across these areas as well as within each of them. A tool that affords an integrated approach, combining data from each portfolio, is powerful in the analysis and examination of information across multiple dimensions.
- Tools that allow for project documentation to be stored within the portfolio and attached directly to projects is a great feature to have as it affords a project manager or team the ease of going to one place for all necessary documentation.
- Many tools offer the ability for advanced operations and functionality like scenario building and execution and Monte Carlo analysis for prioritization.
- A final consideration is the reporting, analysis, and visualization of the data from the tool. After the organization has taken time to input and maintain data in the system, it is important for the data to be used to make decisions and provide leaders with the information needed to run the organization. Many tools come equipped with a robust set of predetermined reports, graphics, and dashboards so

that it's relatively easy to see the results of the data in the system. Still others offer intuitive and easy to use functionality to build and create reports and graphics to meet specific needs. The reporting and graphics often become the face of the PPM initiative, which is why it is such an important consideration in the selection of the tool. Figures 9.6 and 9.7 are some examples of the visualization of data.

• In more mature PPM implementations, data from other systems such as human resource information and financial information need to be integrated with the data collected through the portfolio processes to represent a complete picture for reporting and graphics. A tool that allows for the integration of this type of data certainly elevates the decision-making capability of the organization because it pulls in more dimensions of data. In these types of mature portfolio implementations, a PPM tool becomes a critical enabler in its capability to perform analysis and provide the information that leaders can use to make informed decisions. As a result, this capability can become a strategic and competitive advantage.

As seen in this discussion, technology plays an important role in process foundation, particularly in defining the desired future state. In conjunction with forming the process team, we recommend that a technology team be formed as well. The technology team comprises existing organizational resources and/or consultants who are experts in the selected tool. The technology team should be key participants in the process workshops since the workshops serve as a forum for defining the technology/tool solution requirements. As process is being defined and the overall process map begins to take shape, the requirements for the technology/tool also are taking shape. At the same time, the Technology team should be proactively building prototypes of the process maps, the data entry screens, and related reporting and graphics interfaces. As a series of data entry screens are prototyped to mirror and enable the process and data requirements, the technology team should present them in the workshop to demonstrate to the process team how the process can work. The process team should provide feedback to the technology team about what is working well and where changes should be made. In addition, the process team will have an opportunity to envision how the process will work by interacting with the prototypes and thus may decide to change or enhance a part of the process. These types of interactions and discussion should continue throughout the workshops and prove to be

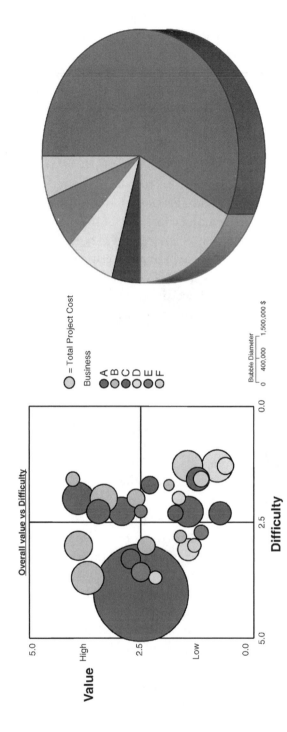

Figure 9.6 PPM tools present data visually to aid decision making. This graphic shows the portfolio of projects in terms of the value they deliver to the organization relative to the difficulty in implementing them.

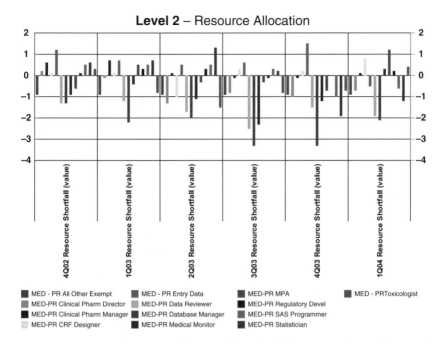

Figure 9.7　PPM tools present data visually to aid decision making. This graphic shows the distribution of resources across the portfolio of projects.

extremely valuable to the process team, the technology team, and the people team. From these iterative design sessions and interactions between the process and technology teams, the process begins to come alive for the team because they begin to see what the desired future state will look like and continue to drive toward alignment with the objectives of the PPM initiative.

TYING IT ALL TOGETHER: PEOPLE, PROCESS, TECHNOLOGY

The people-process-technology triad is paramount to the establishment of a strong foundation on which Project Portfolio Management can achieve successes and mature to the appropriate level the organization needs. Even after the initial implementation of PPM, it is important to keep a balanced approach between people, process, and technology because each area of the triad can be affected by a change in the organization strategy, environment, leadership, or culture. Relying on this balanced and integrated approach has proven to be an effective approach in the evolution and maturation of PPM as a strategic and competitive advantage.

CHAPTER SUMMARY

Foundational Principles

- The successful implementation of PPM is a journey, not a destination.
- The business case for PPM should tie back to the organization's strategy.
- The successful business case for PPM documents the benefits it can bring to the organization.
- PPM success requires a balance of people, process, and technology.
- Engaged and empowered people are keys to PPM success.
- The PPM sponsor should have enough authority to ensure that an organizational environment exists to enable success.
- Establish the process foundation in three steps: define the current state, identify issues and opportunities, and describe the future state.
- Use common, standard terminology in communicating about PPM.
- Alignment of PPM objectives and level of technology is critical.

Foundational Tools

- Organization map
- Organizational assessment
- PPM process flowchart
- PPM software

Lessons Learned

- As organizations improve their PPM processes, project performance improves.
- Stakeholders need to be engaged early and invested in the PPM implementation process.
- Create a standard repository of PPM training materials, including policies and procedures.

Chapter 10

PPM Design

Design is an interesting word. According to the *Merriam-Webster Online Dictionary (Design 2009),* it can function as both a verb and a noun. This fascinating dichotomy actually fits quite well with the idea of Project Portfolio Management design.

Consider first that a verb is deemed an *action* word. *Merriam-Webster* states that a verb "expresses an act, occurrence, or mode of being." The dictionary goes on to state that the word *design,* when used as a verb, means literally "to create, fashion, execute, or construct according to plan: devise, contrive."

Consider also that a noun is the *subject* of a verb. The dictionary states that a noun is an "entity, quality, state, action, or concept." In this application, design is defined by *Merriam-Webster* as: "1a: a particular *purpose* held in view by an individual or group <he has ambitious designs for his son>; b: deliberate purposive planning <more by accident than design>; 2: a mental project or scheme in which means to an end are laid down."

Putting these two applications of the word *design* together points out that the design of PPM is essentially a two-fold undertaking. It is an *action* involving the definition of a company's or organization's *purpose* and application of that purpose through a plan or mental project to bring about a desired state. PPM design is also about *creating*, fashioning and devising a process for use in a company or organization. This is just what we will explore in this chapter, but not strictly from a theoretical perspective. We will share our experiences and lessons learned in the real world when designing PPM—in both its verb and its noun usage.

Foundational Principle

PPM is a two-fold undertaking: (1) action and (2) creating.

PPM's SEVEN Ps

We developed this simple mnemonic as an outline and affirmation of PPM design principles to encourage one another and ourselves. The seven Ps refer to the first letter of key design principles we discovered that when used together make PPM work. The Ps are *Passion, People, Politics, Process, Potential, Performance,* and *Payback.* Let's look at each one in turn.

P^1 = Passion

We all share a real *passion* for the subject of PPM. This passion is what keeps the design alive within us, especially when things may not go as smoothly as we might hope.

Let's go back to the good old dictionary to gain a better understanding of what we mean by passion. We use passion along the specific definition *Merriam-Webster* describes as "a strong liking or desire for or devotion to some activity, object, or concept."

We have found that many of us have our business experience rooted in the discipline of project management. From that base we each moved on to take in the best from other disciplines, including program management, systems engineering, configuration management, financial investment, operations management, and enterprise architecture. Our journey found us each yearning for something more. We sought a better way to put the best of these disciplines together to more effectively and efficiently run our

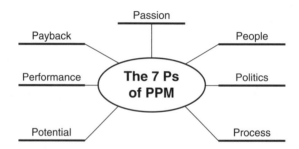

Figure 10.1   The Seven Ps are key design principles that, when used together, make Project Portfolio Management work.

businesses. We each gravitated into the emerging discipline based on a more than half-century-old financial concept applied to the management of projects and programs.

Dr. Harry M. Markowitz, the 1990 Nobel Prize winner in economics, is often called the father of modern portfolio theory based on his 1952 article "Portfolio Selection" published in the *Journal of Finance*. Later, in the 1990s, people such as Robert Cooper, Scott Edgett, and Elko Kleinschmidt (1997) began applying Markowitz's theories to the management of programs and projects. And so began the subject of this book—Project Portfolio Management.

Because of our wide variety of experiences from a multitude of industries ranging from pharmaceuticals to information technology to retail grocery to aerospace, we discovered that PPM became a logical arena into which we could all throw our hats. We have found this arena a great place to integrate the various disciplines noted earlier into an improved way of running a business. In fact, we actually refer to PPM as more of a practice than a discipline.

It's from this experience and practice that our passion has grown. And this passion only grows brighter the more we practice and experience PPM. Not that everything is always coming up roses. Wait, maybe it really is roses—we seem to find the beauty and wonderful fragrance, but also get stuck by a few thorns along the way.

Zander and Zander (2000) put it this way: "The access to passion gives momentum to efforts to build a business plan, it gives a reason to set up working teams, it gives power to settling individual demands, and it gives urgency to communicating across sections of a company."

Foundational Principle

Passion holds PPM together.

So, passion is the first P. It is perhaps the one P that holds all the other Ps together.

P^2 = People

Although we tend to use a lot of numbers in the practice of PPM, it's not really that cut and dried. PPM is first about *people*. It's people who do the work of a project, who make decisions about projects, and ask questions about projects. But people also come to the table with all the idiosyncrasies that combine to make up a unique individual.

Foundational Principle

PPM is first about people.

People are not numbers, and they usually do not want to be told that some calculation about a project is the answer to all their problems and issues. A PPM manager really needs to be a student of people, especially their behavior and motivations. They are the wild card in PPM, but they also make the job challenging and exciting. People keep us all reaching for the stars.

This is quite a contrast from the comments we sometimes hear, such as "This job would be great if only I didn't have to deal with people!" Every so often, all we need is a little attitude adjustment on how we look at people. It is also important to each of us to know that we're not alone and that we've all been there. We all need to cultivate a network of people we can trust and with whom we can comfortably bounce ideas around. The Enterprise Portfolio Management Council (EPMC) is one example of a network of people that fit this mold.

Realizing that so much of our job is about understanding human nature is a big step in the design of PPM. It may seem obvious, but working with people and understanding them is one of those difficult areas to master in PPM, and in life. In fact, it often may seem like more fun to work in the realm of PPM and its numbers, templates, charts, graphs, and computing systems, rather than deal with people. For example, when I went searching for information about dealing with people in project management and PPM, I discovered that the landscape is sparse. There are mountains of information on the tools, templates, computing systems, and procedures of project management and PPM, but not much on the "people side" of the equation.

But here's the one message we hope everyone gets from us loud and clear: PPM is about people. People make it work. People make it fail. People make it frustrating. People make it challenging. But people also make it fun and rewarding. PPM is about people we work with, people we impact, and also about PPM managers and business leaders. After all, PPM managers and business leaders are also people! Keeping this in mind, we can take comfort that we are not alone in trying to get PPM to work. We all are in the same boat—PPM managers, business leaders, stakeholders—everyone. If there is one thing we've learned being a part of the EPMC, it's that we are all in this together and it really doesn't matter what industry we're in; we are all dealing with the same big issues. And that's the beauty of it—we can all learn from each other.

P^3 = Politics

"Ugh!" we may exclaim, "I hate politics." Why can't people just look at the facts and use logic to make decisions and agree with plans? I don't trust all those brown-nosing people at the office. They don't want to get any real work done, other than on their own careers."

We know at the sound of the word *politics* all kinds of sleazy, kowtowing images of smoke-filled back rooms with all the "good old boys" come to mind. But politics is really a good word to describe the essential tenets of this piece of the seven Ps.

What do we mean?

Politics is really about influence, negotiation, power, and autonomy. As Aristotle (2004) said, "man is naturally a political animal." *Merriam-Webster* defines politics as "1a: the art or science of government; b: the art or science concerned with guiding or influencing governmental policy . . . ; 5: the total complex of relations between people living in society."

Foundational Principle

Politics is really about influence, negotiation, power, and autonomy.

The P of people and the P of politics are very closely related. One deals with understanding people and the other deals with understanding the effective methods to work with people.

We've all probably learned when we were project managers that there are all kinds of different sources of power or authority, like boss/subordinate, charismatic, purse-string, bureaucratic, and so on. In the capacity of a project manager we usually had very little direct power or authority over the people we had working on the project. We had to learn how to convince folks to do the various tasks on the project, anyway. We were actually playing politics, even if we didn't call it by that "horrid" name.

Being a PPM manager is much the same; the people working on projects that are part of the portfolio aren't all reporting to us. Consequently, we have to learn how to influence people by using the power or authority available to us by negotiating an agreement while not stepping on the autonomy of others.

Autonomy is really about each of us feeling in control of our lives. Another way of putting it is being able to maintain our feeling of independence while keeping our egos from getting bruised. As PPM managers, we have to be very careful how we treat those around us. If we come on too strong and demanding, then we will have limited ability to influence those folks on later projects, even if they do what we ask this time. In the long run, we will hurt ourselves in being able to do our job.

P^4 = Process

The fourth P is process. It is about creating, using, and following a PPM process.

This is one place where we have seen a lot of good intentions, but not always good results. We thought this would be one of the easiest parts of PPM, but it evidently is not. That being said, it really can work and when it does it's great!

The whole idea behind PPM is to make sure our business is doing *the right work*. Project management processes are how we do the work the right way. Cooper (2000) states these same two concepts as "Doing the right projects" and "Doing projects right." We've seen a number of different PPM processes, but they all boil down to the same basic elements. Some companies like to have few steps at the top; others want to see the flow of a project through the process from beginning to end. One simple process flow we developed at the EPMC has just 10 steps (see Chapter 8).

Lesson Learned

PPM is doing the right work. Project management is doing the work right.

The real key is not so much whether you *have* a process. We've seen all kinds of processes documented down to the gnat's eyebrow, but they weren't being used or followed. And that last part is the real litmus test—is the process being *used and followed*? PPM is all about the *right action*—when it's done right!

Foundational Principle

PPM done right uses and follows a process.

The important thing about any PPM process is not so much the number of steps, but to have specific, understood, and documented decision points *in the process*. What we mean by this is that there needs to be at least four key decision points in the process:

1. The first point is where we decide if we should even pursue an idea and build a business case and project plan.
2. Number two is where we decide if the project should be authorized to go into the portfolio.
3. The third decision point is whether we give the project either the go, no-go, or wait decision.
4. The final decision point is whether we have finished, reaped the benefits, and can close out the project and send it to the project portfolio archives.

There are a couple of things to watch out for in a process design. There may be a tendency to evaluate one project at a time as it goes through the various decision points. Each project needs to be examined and evaluated, but they also need to be put into the context of the portfolio, particularly before they are authorized for implementation. Also, it can become easy to try to simply compare one project to another at each decision point. Using some advanced techniques such as the Analytical Hierarchy Process (AHP) can do this most effectively. Thomas L. Saaty (1994), the originator of AHP, describes it this way: "AHP is about breaking a problem down and then aggregating the solutions of all the subproblems into a conclusion. It facilitates decision making by organizing perceptions, feelings, judgments, and memories into a framework that exhibits the forces that influence a decision."

Lesson Learned

Beware of the tendency to evaluate one project at a time as it goes through the various decision points.

Another key element in PPM is the management of resources. So we need to be vigilant that we are not authorizing projects into the portfolio without performing our due diligence about the availability of resources to work on the projects as they are currently planned.

Succumbing to these pitfalls may result in a suboptimized portfolio. One indicator that this may be happening as a process begins to be used is to measure how many projects are getting a yes vote at each decision point. If every project gets a yes, then there may be cause for concern that the PPM criteria are too vague. The unwanted result of this kind of situation is clogging our pipeline of projects and overburdening our people.

One technique that we have found useful in a PPM process is to introduce a classification system for projects. This is not the same thing as a priority or urgency schema. Rather, it is simply putting projects into categories that allow the process to operate more smoothly and ensure that the right projects rise to the top.

Some suggested classifications are as follows.

- **Must do:** projects that are strategically essential to the operation of the business, or are mandatory to complete. Mandatory projects may be from regulatory requirements levied on the business, or even company directives that mandate action. To determine if a project is really mandatory, ask: "Can we simply not continue to do business without this project being successfully implemented?" In a production environment this may take on the nuance of "stopping the line" and delivery of the product until this project's deliverables are in place. All too often we have seen where "mandatory" actually meant "I really want it!"
- **Need to do:** projects that are important for meeting customer requirements, improving a product in a significant way, or enabling major cost reductions, but are not mandatory in nature. These are the first projects in the discretionary pool.
- **Should do:** projects that fall deeper in the discretionary pool. These are ones that we may choose to do if enough resources are available.

These are projects where there appears to be a moderate improvement in the product or cost savings or avoidance for the company.

- **Could do:** projects that are at the bottom of the discretionary pool. These are projects that are "nice" to do, if we can get to them given all the other demands we are attempting to meet.

P^5 = Potential

The fifth P is potential.

This is the forward-looking part of the process. Ferreting out the right projects using the PPM process does the selection. But there is still the matter of balancing the portfolio.

Foundational Principle

Potential is forward-looking.

We're not going to start talking accounting now, are we? No. Balancing the portfolio means making sure that our strategies and other key criteria are aligned the way we want them given the current environment.

That is understandable. But just how do we do this?

Remember those old vinyl records we had, or at least saw around our parents' or grandparents' house? A preschool teacher was putting an LP (long-playing record) on an old phonograph to play a song for the class. The room of three- to five-year-olds looked at that record and one wee one exclaimed, "Teacher! That's a really big CD!" Oh, the innocence of youth!

The way in which we all use the more sophisticated sound systems that seem to abound in today's culture can be used as an analogy in reaching the optimum potential in a project portfolio. In the "old" days, the really fancy phonographs that played those records had all these switches, dials, or levers we could move around until we got just the right sound. Now most sound systems, CD players, and MP3 players have those sound equalizers built right in. We can do the same kind of thing with our project portfolio.

What's a sound system got to do with balancing a portfolio?

When we set up the PPM process and use the business case, we create the decision criteria for the selection and prioritization of projects. That's not a static thing. If we weight the criteria, then we can let each criteria weighting act as one of the levers or dials on our portfolio equalizer. By doing this we can get just the right balance, or fidelity, we want from our portfolio based on our strategies and the environmental conditions our companies operate within. We can keep an eye on the portfolio, the strategies, and environmental conditions and make adjustments quarterly, annually, or whenever some major event occurs in our business or its environment.

We can move the weighting of each criterion to suit the outcome we want to get from the portfolio. We can determine the weighting of each criterion by looking at what is important right now in the business and even react to crises, if needed. That keeps the portfolio a living and dynamic business tool.

DECISION CRITERIA

The decision criteria are usually of two types used for two different purposes.

Lesson Learned

Criteria are essential for qualification of project proposals and prioritization of projects in a portfolio.

The first type is often called screening or qualification criteria. The purpose of these decision criteria is to screen, or qualify, project proposals. These criteria are used to eliminate projects that do not meet some essential business strategy or set thresholds for investment. For instance, these may include key business absolutes, such as:

- Is this project in "X" market?
- Will this project cost more than $X next year?
- Is this project being done to meet a regulatory requirement?"
- Is this project being done to meet a company directive?
- Does this project exceed "X" level of risk?

These are criteria that can easily be given a yes or no answer based on whether they meet the standard set.

The second criteria type is usually termed prioritization or evaluation criteria. The purpose of these decision criteria is to prioritize, or evaluate, projects that have passed the initial screening, or qualification, hurdles. These criteria are not usually "yes or no" in nature, but are usually measured in ranges of values. In addition, these criteria can be weighted and used to balance the portfolio effectively.

We have found through experience and practice that prioritization criteria tend to fall into four major groupings. These are value, strategy, balance, and linkage. Generally, these groupings are further subdivided into subcriteria to model the concerns and direction of the business.

Under the value grouping there will often be such subcriteria as:

- Cost savings—measured via Net Present Value (NPV), Modified Internal Rate of Return (MIRR), Internal Rate of Return (IRR), breakeven, or payback
- Cost avoidance—measured similarly to cost savings
- Return in (XX) Months—measured in dollars over a set time period

Under the strategy grouping common subcriteria are:

- Support to corporate strategies—qualitative measures such as "strongly or directly supports," "moderately supports," or "does not support"
- Support to business unit strategies—similar qualitative measures as corporate
- Support to Organizational Strategies—similar qualitative measures as corporate

Under the balance grouping common subcriteria include:

- Intangible benefits—qualitative measures on items such as business impact
- Probability of success—inverse of risk measured by such items as complexity and urgency
- Resource availability—measured in terms of timeframe of resource availability

Under the linkage grouping common subcriteria include:

- Indications of how the project under consideration is dependent on other projects or is a trigger, catalyst, or required item for another project to complete; not usually a quantitative or qualitative measure

---

Lesson Learned

Clearly define each criterion, its measures, ranges, and score given for satisfying each range. It works best to do this before activating the PPM process.

SOURCE OF DATA AND INFORMATION RELATED TO DECISION CRITERIA

How do we get the data and information that will make it possible to use the criteria? The best source of this data is directly from the business case. This is where development of standard business case templates and guides can be invaluable in successful implementation and operation of PPM processes. The business case not only should provide data for scoring each of the decision criteria used for prioritizing projects, but it should also explain any assumptions used to reach the values put forth in the business case and explain the various options considered for meeting the project requirements.

Lesson Learned

Link the decision criteria to business case requirements via documentation templates or specific data view completions in computing systems.

Use of the business case and project plans enables the business case creators to do the necessary scenario and option analysis to develop a sound recommendation for review by the decision makers.

SCORING PROJECTS AND PORTFOLIO

Each project should be scored against the criteria used in the prioritizing of the portfolio. This scoring should be done independently. Usually, the creator of the business case and the sponsor will create the first scores. The business leaders often will review and adjust the scoring based on consensus of the group.

We have found it invaluable to document the scores in a matrix, or scorecard. In addition, recording the reasoning behind the score at the time it is done will save time in the future when reviewing the scorecard with the business leaders.

There are numerous acceptable methods for scoring projects. Some scoring is done on paper, but most often it is done in a computerized spreadsheet program. For those businesses fortunate enough to have PPM software available for use, this feature can often be configured in the tool.

Some examples of scoring methods include prioritization using:

- A combined total value of all groupings into a single number
- Subcriterion scored individually, summed by grouping, and then weighted by grouping
- Subcriterion scored individually, and then weighted by individual subcriterion.

Lesson Learned

Start with a simple scoring mechanism. Add more complexity to the scoring after PPM has been in place for a time.

There are multiple methods for scoring the portfolio. This is sometimes referred to as "racking and stacking" the projects. These methods include:

1. Place each individual project into the portfolio and order the projects by total score. This is perhaps the simplest method.
2. Score subcriterion individually, with weighting applied by individual subcriterion, then ordered by contribution of that subcriterion when compared to all other portfolio project's subcriterion.
3. Do the same thing as in method 2 above, but take it one step further by summarizing the subcriterion by grouping, and then calculate the contribution of each project to the overall grouping and order the projects accordingly.

4. Put projects into the portfolio, order them by project classification (for example, "must do," "should do"), and order projects in each classification by either method 1 or 2 above.

Lesson Learned

The mechanical scoring of projects in the portfolio is only the starting point. Numbers are just indications of what the priority of the projects could be. The final decision on prioritization of the projects is up to the business leaders.

WEIGHTING DECISION CRITERIA

Weighting the decision criteria gives flexibility to business leaders in operating the portfolio. Weighting is simply a reflection of the importance of each criterion to the business at any given point in time. The weight assigned to each subcriterion will likely change over time.

Some of the reasons the weighting might change include shifts in strategic direction for the company, business unit, or organization; a change in the business conditions; or fluctuation in the marketplace.

By using the weighting method, business leaders can quickly respond to the various changes noted above by adjusting the various weights to more closely align to the current business situation. This is moving the dials and levers on the portfolio equalizer. Ultimately, though, it allows the business leaders discretion in the final ranking of projects within the portfolio.

DRAWING "THE LINE" IN THE PORTFOLIO

One of the strengths of PPM is the ability to judge and manage resource capacity within the portfolio of projects. Once the projects are prioritized in the portfolio, then the pertinent people, machines, facilities, and so on can be applied to the projects in priority order until the resource pools are exhausted. We have found that it is particularly important to pay attention to key resources used in the execution of projects. These key resources may take the form of a subject matter expert (SME), a computing system manager, enterprise architect, design engineer, or even project manager.

The main point here is that once the resources are exhausted (not literally, we hope), then "the Line" can be drawn on the portfolio of projects. All the projects above the line are given the authorization to implement. All the projects below the line may be approved to be in the portfolio, but since resource constraints now exist, these projects must be put on hold until resources become available. It is also at this time when the project managers, PPM manager, and business leaders can examine the results of the application of resources to the projects as prioritized and see if this really makes good business sense. These folks can then work together to reorder projects as needed and adjust resources according to the final project priorities.

It's at this point in the process where PPM really pays dividends. All too often, before PPM is implemented in an organization, it may have been a roll of the dice as to whether projects would get the necessary resources to meet customer commitments and schedules. Now the project managers and business leaders can actually manage the portfolio of projects and be able to explain to stakeholders, customers, and others in the organization why certain projects may not be progressing as expected.

Lesson Learned

PPM enables effective resource and capacity management.

So now what can we do for those project sponsors, advocates, stakeholders, customers, and project managers who did not see their projects make it "above the line"? This is where the earlier Ps of people and politics really come into play.

We will need to talk with these folks and explain that we understand their need, but that we just cannot do their project at this time. This may be a time to discuss constraints facing these folks and the project itself and ask if things can be adjusted. This could include changing the requirements of time, money, quality, or performance and scope of the project to enable it to move higher in the portfolio. We may be able to assist them in looking for alternative means for accomplishing the project, such as outsourcing or bringing in contract labor.

We've hit on Passion, People, Politics, Process, and Potential. We even snuck in a little about the business case. We still need to go over performance and payback.

$$P^6 = Performance$$

The sixth P is performance.

This is where we review how well our projects are executing according to their plans, including where we are on the cost/benefit graph. We'll talk more about that cost/benefit graph with the next P, payback.

Here it's best to keep things simple and to the point. To tell how the portfolio is doing, we look at the contents of the portfolio—the projects themselves. This is nothing more than tracking project status and giving periodic reviews to the decision makers. This is also where consideration is given to cost, schedule, quality of deliverables, and scope of the projects. Sometimes the project is at a point where a key decision needs to be made on continuing as is, making some changes, or killing the project. These are the tough decisions.

Foundational Principle

The sum of the performance of the projects tells us how the portfolio is doing.

This brings us to a point we kind of glossed over. The decision-making group that heads up the portfolio can be called by any number of names—portfolio team, executive team, leadership team, decision board, or portfolio management board, to name just a few. It really isn't so much the name as the people who make up the board that is important. We have found that it is best to have the real decision makers who have the people report to them that work on the projects and the ones who hold the purse strings in the organization to be board members. This is essential to the successful functioning of PPM overall. We want these people to be the ones having the discussions and making the decisions. Our jobs as PPM managers are to bring the board the data and recommendations then let them discuss and decide.

This is where PPM has the rubber meet the road, so to speak. The conversations about the projects, the portfolio, and the organization are paramount. It is through these discussions that understanding is gained by business leaders, PPM managers, project managers, and stakeholders. Out of understanding come informed decisions. This is one of the positive outcomes of PPM—making good decisions to enable the organization to operate more effectively and efficiently to meet its strategies.

Lesson Learned

Good PPM provides the forum for valuable discussion, which enables informed decision making.

P^7 = Payback

The final P is about payback.

Payback, as it is used here, is not about revenge. Rather, it is about projects making good on the cost/benefit presented when they were authorized to go forward and implement.

Foundational Principle

Payback is the real bottom line of PPM.

A simple way to check this out is with a payback graph containing four tracking lines on two axes.

- On the horizontal axis is time.
- On the vertical axis is dollars.
- One line shows the expected expenditure rate from the top down; that is, using the total amount budgeted at the far left and decrementing it each time period as budget is expected to be used.
- A second line that follows this one presents the actuals for the same data elements.
- The other two lines also start on the left, but they begin at zero.
- This second set of lines track the planned and actual benefits as the project progresses.
- The payback point is where the two lines intersect.

The payback graph is used for a couple of different purposes. One use is just to see how the project itself is doing against the expectations

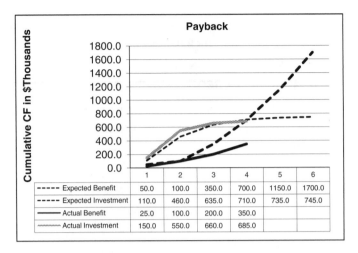

Figure 10.2   Payback graphs are used to show whether projects have made good on the cost/benefit presented when they were authorized to go forward and implement.

generated by the business case. The second use is for the portfolio. In this situation, the decision makers use the chart to determine continuation of the project when the portfolio is reviewed. It may be that the project will be allowed to continue when others are put on hold or cancelled because the payback, or breakeven, point is near.

Lesson Learned

Simple visual tracking of payback progress works best.

That wraps up the seven Ps. We alluded to the business case a few more times, but not the design concept specifically. Now let's finish this chapter.

LINK TO THE BUSINESS CASE

A business case is far more than just a cost/benefit analysis (CBA). Unfortunately, the term "business case" has been used so often that people have begun to equate it with a CBA only. But done right, the business case becomes a key opportunity for the leadership to discuss where the business is going and how the particular project being described in the business case fits into the project portfolio.

Foundational Principle

The business case is more than just financial analysis.

The business case is one of three essential deliverables to the decision board from the project manager and sponsor. The other two are the project proposal and the project plan. Hey, another set of Ps! Yes, that's true, but these Ps are not part of the PPM Design.

Let's finish up here. As we said, the three deliverables in the PPM process are the project proposal, business case and project plan. Each requires a different amount of effort, but all three have some key elements in common. Each one addresses the decision criteria set forth in the PPM process, just from a different perspective.

- The proposal is just a high-level, rough-order-of-magnitude look at a project idea.
- The business case and project plan are the detailed analysis done to address the decision criteria for approval to be part of the portfolio and implementation.

BUSINESS CASE

The business case consists of several major elements. Often the first element brought up when thinking about a business case is the financial model, yet the financial model is not all there is to a business case.

Often the terms "business case" and "financial case" are used interchangeably. This may be the result of many people's emphasis on the financial elements, or models, involved in the business case. However, a business case considers more than just the financial aspects of the decision under consideration. Some folks will think they require a business case, but really need a project plan, business plan, or operating budget to address the given situation. All too often, people will bring forward a "business case" for consideration, when what they really are supplying is a sales presentation for a specific idea that explores only one option (Stratton, 2004).

The business case should present clear evidence and reasoning that supports the conclusions presented in terms that can be understood from a business, technical, and financial perspective. A key component in any business case is the financial model constructed to simulate the business problem or opportunity, as well as the expected results of following the ultimate recommendation (Stratton, 2004).

Why Do a Business Case?

A business case is in order any time there is a decision to be made regarding an investment of the company's money either through distribution of capital or use of other company resources, including labor. Documenting the reasons for the investment, the options available, and describing how the investment helps the company reach its goals goes a long way toward obtaining the necessary decision and funding (Stratton, 2004).

Developing a Business Case

"Developing a business case is really best done by following a simple, systematic process," states Stratton (2004). "The development of a business case as a process which can use multiple tools does not equate to being a bureaucratic and time-consuming endeavor. A business case developer can move through all the process steps and use the tools quite quickly and efficiently, especially if they are well-versed in the mechanics of the tools and process steps."

Suggested business case process steps are shown in Figure 10.3.

Financial Aspects of a Business Case

Certainly the financial aspects of a business case cannot be ignored, but neither should they be considered as the sole data from which to make a decision (Stratton, 2004).

The key to a successful financial analysis is the construction of a sound cost/benefit model, also called an ROI (Return on Investment) model. The ROI model is usually a template done in a spreadsheet format. There should be either a worksheet for each option, or a view of all options together in the model. The model usually has two distinct parts: one that builds the values based on the specific variables and factors pertinent to the business case and the other that summarizes the data into a standard format for ease in comparing options and supplying the necessary data for the decision makers. The gathering of the correct data is essential in producing a plausible and acceptable business case (Stratton, 2004).

Business Case Process

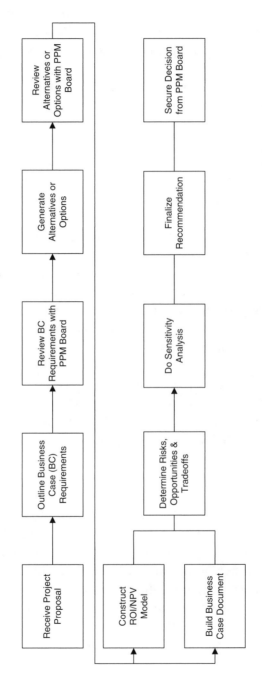

Figure 10.3 Developing a business case is best done by following a simple, systematic process.

The ROI Model is a separate spreadsheet done in a computer program such as Microsoft® Excel® and may be available in multiple template formats depending upon the complexity of the financial analysis required. There are primarily two types of models that can be developed:

- Deterministic: This model identifies variables, but "determines" a single outcome value for each one. The values are determined using assumptions about the variables. Assumptions may be established through "what if" techniques, or by specific predictions based on experience or expertise. The combinations of assumptions for all the variables in the model are referred to as scenarios. Each scenario will produce a single outcome.
- Stochastic (pronounced "stow-kastic"): This model identifies the variables that are uncertain and uses statistical methods to produce a range of values. This range of values uses probability distributions in establishing the limits of those values. These probability distributions model the understanding of risk and uncertainty about the variable's possible values. These stochastic variables can also be referred to as random, chance, probabilistic, or uncertain quantity variables (Stratton, 2004).

The combinations of these variables and their probability distributions will produce forecasts of possible outcomes based on a selected number of trials when using Monte Carlo analysis. The forecasts will give a range of values along a confidence interval (e.g., 80 percent confident the value will be $X or greater) (Stratton, 2004).

The financial model should include the following:

- Financial measures: ROI, Net Present Value (NPV), Internal Rate of Return (IRR), Modified Internal Rate of Return (MIRR), before-tax net cash flow, and after-tax net cash flow. These should be totaled in a specific location on the spreadsheet, preferably at the top.
- Major financial headings: Benefits/Gains, Operating Expense Items (including various forms of labor), Capital Assets Purchased, Cash Flow Summary (including depreciation as applicable).
- Graphical charts: annual net cash flow, cumulative net cash flow, payback.
- Key metric chart updated throughout the lifecycle of the investment: total cumulative planned investment compared over the same time

period with total cumulative planned return; and total cumulative actual investment compared over the same time period with total cumulative planned return (see Figure 10.2). This chart works best when it shows both a table of values and a graphical representation of the data (Stratton, 2004).

This model will be driven by the isolation of the necessary data that drive the business. This allows correlations to be established, estimates done, and values calculated. The financial model will aid in a key part of the business case analysis, that of the sensitivity analysis. A sensitivity analysis, in a nutshell, asks:

- What variables (values) can change the option's outcome?
- Are these values reasonable? (Stratton, 2004).

This analysis reveals the variables that contribute the most to the final outcome. This information is important in the decision being made for the business case. The analysis provides insight as to the impact of these key variables on the value of the options under consideration in the business case. This analysis tests limits, strengths, and weaknesses. It also identifies threats and opportunities, calculates ranges of possible outcomes, and provides other insights (Stratton, 2004).

The goal of this analysis is to provide decision makers with the facts, data, and analysis required to make an informed decision. The decision support tool also should clearly delineate the tradeoffs in making the decisions associated with the business case. Completing this analysis will aid the decision makers in concentrating on the right issues surrounding the business case and not waste time "majoring on the minors" (Stratton, 2004).

Business Case Document Components

The business case document is the artifact that collects the facts and data surrounding the problem to be solved or opportunity from which the company may benefit. The document becomes useful in the implementation of the project, or program, by providing the necessary scope of the work and its ultimate purpose (Stratton, 2004).

This business case document should explain why the funds are needed, and provide sufficient information to help weigh this requirement against other needs (or investment proposals) that are competing for the same funds and other resources. The business case also should explain the root causes or drivers of the situation, list the stakeholders for this proposal, and itemize relevant environmental considerations or factors (Stratton, 2004).

The document should show the business value of the proposed investment by examination of the facts and data that are both tangible (often financial) and intangible. In writing the document, be sure to make your conclusions explicit; *don't assume* everyone will be able to draw a conclusion, or will draw the same conclusions you did. Finish strong; close with a specific recommendation, especially if the case involves a funding request. Make it very clear that "the ball is now in the decision makers' court." Then assure them that the progress of both the investment and the benefit noted in the proposal being considered by the company will be measured. Be sure to show the decision makers how the measures will be communicated back to them (Stratton, 2004).

The document contains eight major sections. Each major section includes many subheadings that guide the business case developers through a logical progression of questions, answers, and thought by using the tools noted elsewhere in this paper. The following major sections should make up the business case document:

- Executive summary
- Business case overview
- Current situation
- Assessment of options
- Sensitivity and risk analysis
- Contingencies and dependencies
- Recommendations and conclusions
- Metrics (Stratton, 2004)

LINK TO THE PORTFOLIO

The final component of the PPM design is the portfolio itself. The portfolio is really just a concept. So we need a model to represent it. This is where a tool comes in handy. The tool can be paper representing the projects and the output of their deliverables on a visibility wall, or it can be a spreadsheet, database, or specialty Commercial off-the-Shelf software (COTS), or even a Software as a Service (SAS) product. The point is to have a way of putting all the projects together in one place so we can compare the various criteria we have on the portfolio equalizer.

This is where we can now really discuss the *design* concept. Design of PPM is just what we've been discussing in this chapter. Design is taking all seven Ps, putting the puzzle pieces into the frame of a specific business, and linking them together to see the real picture.

The portfolio tool, the three deliverables plus the Seven Ps integrated and applied to an organization, make up the *design* of PPM. That is, every company decides how to develop the portfolio tool, each deliverable, and puzzle piece (or P). It is then in the linking, or integrating, of the portfolio tool, the deliverables, and the puzzle pieces, and applying their use into the company that will reveal a design of PPM specifically for that organization!

Lesson Learned

PPM process designs may vary, since organizations may vary.

A FEW PARTING THOUGHTS

PPM is a dynamic and active process. As with many processes, it probably is not perfect. We haven't found the perfect PPM process yet. If we did, then they probably would not let us work in the place where it is used after all, it's perfect.

Since the process is probably not perfect, we will likely need to make modifications or changes to improve it. The difficult thing here is for us to restrain ourselves and only make changes or modifications if they are truly needed. Likewise the criteria, the weights, scoring, and other aspects of the process may not be perfect. But, again, we need to use restraint and only make truly necessary changes.

Lesson Learned

Processes are usually not perfect, but any PPM process is probably better than no process at all.

CHAPTER SUMMARY

Foundational Principles

- PPM is a two-fold undertaking (1) action and (2) creating.
- PPM is first about people.
- Politics is really about influence, negotiation, power, and autonomy.
- Passion holds PPM together.
- PPM done right uses and follows a process.
- Potential is forward-looking.
- Payback is the real bottom line of PPM.
- The business case is more than just financial analysis.

Lessons Learned

- PPM is doing the right work. Project management is doing the work right.
- Beware of the tendency to evaluate one project at a time as it goes through the various decision points.
- Criteria are essential for qualification of project proposals and prioritization of projects in a portfolio.
- Clearly define each criterion, its measures, ranges, and score given for satisfying each range. It works best to do this before activating the PPM process.
- Link the decision criteria to business case requirements via documentation templates or specific data view completions in computing systems.
- Start with a simple scoring mechanism. Add more complexity to the scoring after PPM has been in place for a time.
- The mechanical scoring of projects in the portfolio is only the starting point. Numbers are just indications of what the priority of the projects *could be*. The final decision on prioritization of the projects is up to the business leaders.
- PPM enables effective resource and capacity management.
- The sum of the performance of the projects tells us how the portfolio is doing.
- Good PPM provides the forum for valuable discussion that enables informed decision making.
- Simple visual tracking of payback progress works best.
- PPM process designs may vary, since businesses may vary.
- Processes are not usually perfect, but any PPM process is probably better than no process at all.

Chapter 11

Implementing PPM

All functioning organizations have PPM in some shape or form. If you look closely, all companies prioritize projects and assign resources to projects for what they think will yield the best returns. This chapter will discuss the maturing of PPM either by improving the existing PPM components or by implementing new components to complement the existing ones to arrive at a more efficient system.

So far in this book you've read about all the elements that make up project portfolio management. You've read about the value and the process in developing a business case and the considerations for PPM design. It's during the implementation phase that the rubber meets the road. The techniques in implementing sustainable PPM are dependent on various factors, such as size of the company, organizational structure, type of incorporation, and so on. Please remember that there is no one standard way to implement PPM. Implementation is a combination of art and science. The PPM design and the development of supporting templates and tools constitute the science portion of the equation. As mentioned earlier in this book, at the end of the day you cannot push PPM without the people who participate in the process. You facilitate the process by effectively managing people.

In the United States, public companies are governed by laws such as Sarbanes-Oxley to be more transparent and to hold officers more accountable than their peers in private or mutual companies. Shareholders of public companies have a right to know how the officers of companies that have their investments make appropriate decisions for yielding a maximum return on those investments. Are the resources of the company used

165

optimally to deliver the greatest return? In this kind of environment, the motivation to implement PPM is external to the organization and hence it could be expected to be driven from the top officers of the organization. The Clinger-Cohen Act requires that government IT investments "reflect a portfolio management approach where decisions on whether to invest are based on potential return, and decisions to terminate or make additional investments are based on performance much like an investment broker is measured and rewarded based on managing risk and achieving results" (Department of Defense, 2000).

In contrast, a private company is not obligated to be accountable to any outside entity. Hence you can expect the decision to implement PPM could come from either someone at the leadership level who is very passionate about it (top down) or from the rank and file of the organization who want to bring some order to their work environment (bottom up). The most desirable state is a combination of both a top-down and bottom-up approach to gain the most synergy among the PPM stakeholders. Also, this approach will provide the quickest, smoothest implementation.

Foundational Principle

PPM is best implemented with a combination top-down, bottom-up approach.

Let's look at the people, process, and technology aspects of an implementation.

EXECUTIVE SPONSORSHIP

No matter what strategy you undertake to implement the process, the importance of having a strong executive sponsor cannot be emphasized enough. There is a difference between having an executive sponsor and having a "strong" executive sponsor. You want this person to *believe in the process* and be passionate about its implementation. He or she should be able to influence peers and be an active advocate of the process. This requires communicating passion to the PPM stakeholders in the organization by creating conversations about the value of the implementation. Conversations are the seeds of change. It is very important to keep the conversation alive, to build the buzz that, in turn, begins to build momentum of the work. A very active, passionate, and involved sponsor can be extremely valuable in keeping the conversations alive. "What doesn't gets talked about, doesn't get done" (Ford & Ford, 2002).

Foundational Principle

A strong sponsor is key to PPM implementation success.

CHANGE

As mentioned earlier, the exercise of implementing PPM is essentially a reengineering project. Reengineering implies "change". We all know how difficult change initiatives are. Change is mostly perceived as being driven by management. If the purpose of the change and the anticipated benefits are not clearly articulated, you begin to experience resistance. So what does resistance look like? The symptoms are unanticipated delays and costs, and stakeholders' just providing lip-service and not complying with the process. You'll find individuals who engage in protracted arguments or try to change the process into something that's more acceptable to them. Or they might totally disregard the process, thus undermining the implementation. The entire experience will have a negative connotation to it. These symptoms will begin to trigger the resistance to change.

Psychologist Kurt Lewin (Chew et al., 2006) offers one of the most highly regarded descriptions of organizational change, which consists of three stages: unfreeze, move, and refreeze.

- Unfreeze: Show the need for change. Minimize the barriers to change.
- Move: Actual change occurs. Maximize the opportunities to change.
- Refreeze: Crystallize new ideas and recognize the acceptance of change.

Throughout the change, managers need to ensure that accurate dissemination and sharing of relevant information is consistent throughout the organization. It means carrying out change by engaging key individuals within the organization and gradually engaging more employees.

Employees should always be aware of the proposed changes because they are the ones who live with the results daily. Managers need to collect feedback about the change continuously by interacting with employees, and then act appropriately to insure acceptance of the change.

SKILLS

Change agents implementing PPM need certain skills. Employees are unlikely to move out of their comfort zone and give up their secure position to adopt new processes and responsibilities unless the change agent exudes self-confidence, has strong convictions, and can articulate a clear vision of the end goal. In short, the person heading up the implementation of PPM must a leader. This person must be good at negotiating and be an effective consensus builder. He or she will have to choose battles smartly, make appropriate compromises when needed, but take a stand on positions when they're right. The person must be business (and politically) savvy, and excellent at influencing individuals over whom he or she has no authority. Otherwise a smooth implementation will be difficult. The heart of PPM is embedded in financial and business decisions. So in addition to all the skills mentioned above, this person needs to be able to talk the language of business and put issues concerning the PPM implementation in terms that business leaders can understand.

Foundational Principle

The person heading up the PPM implementation must be a leader.

STRUCTURE

A number of players support the structure of PPM. Before we talk about individual roles, let's begin with a discussion about some key committees that have to be established. Then we'll discuss the roles and responsibilities of the individuals in these committees. Please don't assume that the committees and roles described below are the only way or even the best way to structure the PPM system. This is just a common way of doing it. Organizations should adapt a structure that is most suitable for their environment.

EXECUTIVE STEERING COMMITTEE

This committee is made up of the C-level executives. They set the strategic vision and objectives for a given program or project. They lead efforts to build consensus throughout the organization to support the project or program objectives. They are also the tie breakers when there's an impasse at a lower-level committee. Ideally, they meet once a quarter to review the status of major projects and programs and also to ratify the priority of ongoing projects as well as any that are in the pipeline or planning stages. The threshold for a project to show up on the executive steering committee's project list is dependent on cost and/or strategic importance of the project as well as the committee members' appetite for detail. There are no industry standards in defining this threshold. Some organizations focus exclusively on capital expense to define the threshold. Some organizations' primary expense "currency" is internal labor cost. You define the threshold that is most appropriate for your organization. For organizations that use the internal labor cost as the primary driver of the cost of a project, we recommend that you establish one blended labor cost for the organization. This simplifies the process of building business cases for the project teams as well.

GOVERNANCE BOARD (DECISION REVIEW BOARD)

This is a formal team of executives from across the organization that ensures projects will meet/are meeting enterprise goals. This group functions more at a tactical level. They have the formal authority to prioritize projects and also approve projects that cut across multiple functional departments. Essentially, it's best if each business unit retains its autonomous right to run its business with minimal external oversight. However, the difficulty comes when different business units want to dip into a common pool of shared resources to execute their projects. This is where the governance board comes into play—to help determine the priority of projects for optimal utilization of resources and greatest returns. Prioritization at this level helps the business units that share resources to better prioritize their work, leaving politics and guesswork at the door.

PROJECT MANAGEMENT OFFICE

The Project Management Office (PMO) is a functional unit that is assigned various responsibilities related to the coordination and management of those programs/projects under its domain. The PMO is also responsible for reporting on the metrics associated with all projects. The scope of the projects PMO tracks is usually dependent on where the PMO reports. The PMO could be within IT, a business unit, or at the enterprise level. There could be multiple PMOs across the organization with a dotted-line relationship to an enterprise PMO. No matter where the PMO resides, a core function of the PMO is to facilitate the process to provide timely, accurate, and credible project information to leadership so they can make informed decisions in a timely manner.

PROJECT MANAGEMENT STANDARDS COMMITTEE

The primary purpose of this group is to develop a common definition of terms and processes for the company that is agreed upon mutually by all business units. The PMO in essence could define all the processes, provide specific definition of all terms, and hand these out to the organization. However, when the PMO does that, in essence acting as the "process police," the chance of getting organizational buy-in is low. There are advantages, however, in developing processes and definitions collaboratively:

- People feel empowered to contribute toward something that has a lot of visibility in the organization and they also feel good about their contribution.
- These individuals become the advocates of a process they've defined. Chances of sustaining the process are higher.
- Communication of the process using consistent terminology across the company is easier, since these individuals share the PM standards committee decisions with their respective departments.

Several other key participants are integral to making the process work: program and project managers, business analysts, developers, and other contributors. Unless everyone performs their roles as per the defined process, PPM will not function optimally.

Figure 11.1 visually represents the relationship among each of the committees described above.

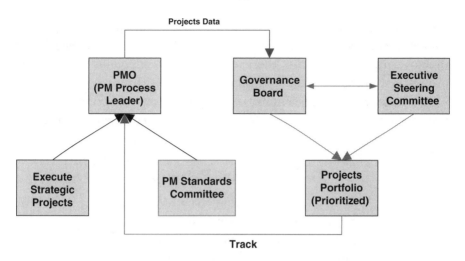

Figure 11.1   Various organizational structures are involved in the execution of the project portfolio. This graphic shows some common ones and their relationships with one another.

PROCESS APPROACH

Now that we've defined a few high-level committees and described their roles, we have to establish how they function and interact. The underlying reason for their existence is to ensure that resources are deployed optimally for the greatest return, working on the projects aligned to the corporate goals. To start off, it's perhaps a good idea to take stock of the inventory of "work." We use the term "work" and not "projects" because, unless you're able to account for all of the work that is occurring in the company and have an inventory of all the resources available, it will be difficult to match your capacity (resources) to the demand (work).

Stage 1

To get started on this daunting journey, it's best to take baby steps, letting the PPM system slowly mature by giving each stage time to stabilize before moving on to the next. So, first have the sponsor of the PPM create a PPM process recommendation committee. Have this committee be chaired by a senior executive who is eloquent as well as an advocate of the PPM process. The committee should be composed of key leaders from IT, profit centers, service centers, and the PMO. Have the committee work together as a group to develop a PPM process framework. Also, have this committee recommend the composition of the governance committee, the executive steering committee, and the project management standards committee. Have this group set expectations for each of these committees. After the process framework has been documented, have this group present their recommendations to the rest of the organization's leadership to get the consensus.

Foundational Principle

The executive chairing the PPM process recommendation committee must be passionate about PPM, understand its value, and have a vested interest in the final outcome.

Lesson Learned

Gain stakeholder consensus prior to presenting the PPM process formally to the executive team.

The benefit of going through this process is that it's easier to get buy-in from senior leadership, which is so crucial for the success of the implementation. Note that neither the project portfolio managers nor the PMO is taking an active role at this stage. The PPM project manager is working quietly behind the scenes to steer the PPM recommendation group in its appropriate direction while actively influencing the naysayers. It's important that you create an environment where it doesn't appear that *any one person* is driving the organization in a certain direction. There needs to be a feeling of teamwork and self-achievement by the PPM recommendation team members.

Foundational Principle

The process design and implementation should not be perceived as any one individual's crusade.

Once the framework is blessed by the senior executives (most of whom have in one way or another contributed to the final framework), it's time to engage the members of various committees identified by the PPM recommendation team. It's best if the PPM project manager and the chair of the PPM recommendation team meet with each of them individually and explain the purpose of their role and the importance of it. It's also important to have these individuals realize that the process has been blessed by the executives at the highest levels of the organization.

The desired state at the conclusion of this stage is that the entire leadership is on board with the PPM process framework and the various committees are in place. The PPM project manager has been given an implicit go-ahead with the entire leadership behind him or her.

Stage 2

Identify the best possible meeting facilitator. Have this person facilitate the project management standards committee and the governance committee meetings. Kick off the PM standards committee meetings and have the committee agree on the following:

- The committee should identify the main categories of work such as projects, maintenance, operations, and the like. Once the categories of work have been identified, very clearly document the definitions of the categories. For example, a project could be defined as "Any work that has a specific start and end date and utilizes resources from departments other than the sponsor's department, and has a total cost of at least $100,000 (inclusive of internal labor cost at the blended rate of $x/hr)." The threshold for the funding of the project is built into the definition of the project. For certain companies, the currency for budget is internal labor cost and not necessarily capital expense. Organizations should articulate the definitions of a project that suits their environment and needs. Given the definition of a project as stated above, any work that falls below this threshold is noncritical work.
- There should be just four or five of the most basic elements of a project tracked consistently across the organization (all departments). Clearly define these elements, for example:

- State of the project (Not started/In progress/Cancelled/On hold/ Closed)
- Milestone status (In concept/Requirements/In IT development)
- Health of the project (Red/Green/Yellow)

Once the committee has agreed on these items, go about documenting an inventory of projects as defined by the PM standards committee.

Foundational Tool

Project inventory.

It's very important not to get overzealous and attempt to document the smallest detail of every project. It is a maturing process. Perhaps the first round is just to get an inventory of "projects" in an Excel spreadsheet. Once you have the list of projects, you could then add a column and write the name of the person who is accountable for the success of the project. If the maturity level of the organization is low, having someone's name that is playing the role of a project manager could be a very challenging task. You then gradually begin to add columns and collect more information. At every stage, display the document prominently, perhaps on your Intranet. The visibility of the document will help to build credibility for PPM. Figure 11.2 is an example of a projects-tracking spreadsheet.

Lesson Learned

Define and document all terms—projects, maintenance, system request, and so on— very clearly. Don't assume that everyone has the same definition in mind.

Stage 3

Once a credible inventory of projects has been established, the governance committee can begin to meet. Though at this stage there may not be formal business case or project charter documents, the individuals in the governance committee can begin to prioritize projects based on their

PPO	Project	Ranking Category	Status	Status Code	PM	Start Date Actual	Estimated	End Date Baseline (Revised Date)	Actual	NPV (60 Months @ 6% Discounted Rate)	Payback Period
Sponsor - A	Project - 1	5	**Yellow**	3	PM - A	04/06/05		12/31/06			
	Project - 2	DP	**Completed**		PM - C	05/01/05	12/31/06	12/31/05 / 01/18/06	01/18/06		
Sponsor - B	Project - 3	1	**On Hold**	2	PM - A	10/05/05	12/31/06				
	Project - 4	2	**Canceled**	3	PM - B	06/01/06	12/31/06	12/31/06			
	Project - 6	NR	**To Start**								
Sponsor - C	Project - 7	DP	**Completed**		PM - B			09/01/06	09/01/06		
	Project - 8	NR	**Green**		PM - E	05/15/06	12/31/07			$137,252	Year 3
	Project - 9	NR	**Discovery**		PM - B						
	Maintenance	N/A	**Green**		PM - E		Ongoing				

Status Key

Green	Project is progressing as planned. Milestones are being met or a feasible catch-up plan exists. Open issues have time for resolution without jeopardizing the Milestone dates.
Yellow	**Two Consecutive** milestones have been missed. Meeting upcoming milestones is in jeopardy. Open issues are <u>not</u> being resolved in the required timeframe. These open issues may slow or change work that is currently underway.
Red	Milestone(s) are being missed. Work activities cannot be performed because important issues are unresolved.

Red / Yellow Status Codes

1	Change in Scope
2	Resource Constraints
3	Change in Priority
4	Date Change by Project Sponsor / Product Manager
5	Technology Issue
6	Vendor Related Issue
7	Risk Mitigation
8	More complex or larger scope than initial assessment

Ranking Order Key

5	Mandatory Projects	NR	Not Ranked
4	Highly Critical	N/A	Not Applicable
3	Very Important	DP	Department Project (Tracked within the department)
2	Important Projects		
1	Will Do As Time		

Figure 11.2   Project information can be displayed in a variety of formats, but it's critically important that information is displayed prominently so all stakeholders can clearly understand the status of the project portfolio.

knowledge of the business and the value of the projects. Why prioritize? There may not be a lot of value to the profit centers to prioritize projects at this stage. However, it's of immense value to the shared services. Shared services serve multiple profit centers or business units. Prioritizing projects at the corporate level makes it easier for the service centers to allocate resources appropriately, removing the guesswork and politicking from the process.

It's to be fully expected that there will be departments that will not play by the rules and will not volunteer project information easily. In such circumstances the service centers can play a very critical role in making sure that everyone complies with the process. In essence, service centers can state that they will not be able to allocate resources to projects that are not on the projects list and have not been prioritized by the governance committee. If you can influence the service centers to abide by this strategy, it will not only help service centers to manage their resources optimally, but will also help mature the process by getting all the departments into compliance.

Lesson Learned

Shared services play a critical role in maturing the PPM process by not allocating their resources to projects unless they're prioritized by the governance committee.

Stage 4

Once you begin to have a stable document that lists all ongoing projects, it's time to begin developing the discipline of building business cases and charter documents. To get the processing going, have the project management standards committee set simple standards for the amount of detail that goes into business cases and charters, making it commensurate with the size and complexity of the project. If it's a small project, perhaps just a few lines or a paragraph will be sufficient to detail the costs and benefits associated with the project. Larger projects will be accompanied by more detailed and thorough charter documents and business cases. The business cases and charter documents will begin to provide more objective information to assist the governance committee in prioritizing projects.

Prior to this stage, prioritization of projects didn't mean a lot to the profit centers. Now, with business cases in place, profit centers can begin to make more informed decisions on which projects to fund and which ones not to. From a profit center's perspective, a business case and a charter document will validate the idea, justify the expense, and develop an understanding of the return on investment.

This is also the best time to introduce the project funding process. The governance committee should work closely with the financial arm of the company to define and set guidelines for funding and monitoring project budgets.

It takes several months for a mid-sized organization to reach this stage. One has to understand that this is a maturing process and that it takes time to build the discipline. Now what? We have an Excel spreadsheet with a list of active projects. Each project on the list probably also documents basic details such as the name of the project manager, status (red/green/yellow), start and end dates, and so on. The projects also probably have documented charters and business cases. Please note that we have not talked about an enterprise PPM tool yet. We have been managing the process with spreadsheets.

Lesson Learned

Be patient in the early stages of the process. Don't get carried away with artificial deadlines and forget the people aspect of the puzzle. Listen to the concerns of the associates and continually tweak the process, not letting it get in the way of productivity.

Stage 5

The various committees that have been established can continue to function, but in a limited manner. Projects can be evaluated, funded, approved/disapproved. We still cannot efficiently forecast resource bottlenecks, cross-project dependencies, or perform what-if scenarios of the impact on projects due to constantly changing business conditions. You will need a robust enterprise PPM tool for this kind of sophisticated tracking and analysis of project data. It's recommended that while the PPM process is maturing, the PMO work closely with the project management standards committee to select the enterprise PPM tool. It's best that these initiatives run more or less in parallel. The PPM tool selection process, implementation, and training of associates on its use are outside the scope of this book. It's also assumed that a well-structured estimation process, resource allocation/management processes, and time tracking would be addressed within the scope of the PPM tool implementation. At this stage, we'll assume that the tool is in place and that all the users have been trained on all aspects of the tool.

The next stage is to mature the process of matching the demand with the capacity of the organization. There are many different variations on the implementation of this process.

CAPACITY

Your PPM tool needs to be able to account for the following regarding the capacity of the organization to deliver the projects in the portfolio:

- All key resources associated with projects and any nonstandard hours that any of the associates work.
- Company holidays (not vacations of associates).
- Block X percent of the availability of the resources across the board for nonproductive time. This time accounts for coffee breaks, water cooler conversations, and so on. Block anywhere in the range of 10–30 percent toward nonproductive time.
- From experience and historical data (if any), estimate the block of time to set aside for production maintenance activities. In the tool, block this time away for all relevant resources/roles.
- Block time away in the tool for resources for any other categories of work the project management committee might have identified.
- The remaining is the capacity available for projects.
- Track vacations, sick time, personal time, and other ad-hoc times-off as and when they occur, to more accurately predict the capacity of resources.

DEMAND

The PPM process as described above, if working well, will paint a picture of the demand. The individual business units will prioritize projects specific to their departments and the governance committee will prioritize those projects that utilize resources from outside of the department where the project originated.

The PMO at a tactical level will facilitate the reporting of all the relevant project data such as business cases, charter documents, cross-project dependencies, resource constraints, and project statuses to the governance committee to help them make informed business decisions.

Figure 11.3 provides a graphical snapshot of the equation between demand and capacity.

COMMUNICATION PLAN

Developing a communication plan for the PPM process is no different than developing a communication plan for a large project.

Foundational Tool

PPM implementation communication plan

Agendas and minutes of all the committee meetings should be unrestricted and accessible to anyone in the company. All the reports generated by the PMO to monitor the process should also be made available to everyone. This information, however, should be in a "pull" mode. By this we mean that the most appropriate way to share this information would be by posting these documents either to a shared drive that is accessible to everyone or on an Intranet website. It would be inappropriate to send an email blast (push) to everyone in the company with the reports and other documents.

Making the information transparent will build trust in the organization. It will also contribute toward delegitimizing pet projects that bypass the scrutiny of the organization.

At a minimum, develop a communication plan to meet the following needs (probably a page on the Intranet):

- Status of projects
- Resource availability reports
- Glossary of commonly used terms
- Other useful statistics

All Projects

Dashboard | Project Requests | Capacity Planner | Projects | Tasks | Reports | Executive | See All ▾

More Views ▾   Export To Excel   Page Options ▾

Save All Changes   Cancel All Changes

Select Resource Types

PROJECT

Project Name	Labor Cost	E Work	Project Type	Milestone Status	Nov 08	Dec 08	Jan 09	Feb 09	Mar 09	Apr 09	May 09	Jun 09
	$3,800.00	95	PLPC Project	Concept / Discovery								
	$1,332,000.00	33,300	PLPC Project	Ongoing	2,779	2,779						
IT MF Dev Mansfield	$57,200.00	1,430			130	130	1,209	1,209				
IT PL GW Dev	$337,920.00	8,448			768	768						
IT PL MF	$154,880.00	3,872			352	352						
IT PMS	$81,920.00	2,048			256	256						
IT Prod Control	$23,040.00	576			64	64						
PL BA	$208,320.00	5,208			372	372	372	372	305			
PL QA Tester	$468,720.00	11,718			837	837	837	837	2,64			
	$11,640.00	291	PLPC Project	Concept / Discovery		20	100	110	60	1		

ROLES

EFFORT ESTIMATE

Show requirements as Hours ▾ per Month ▾

Net Availability

Show as Table ▾

AVAILABLE CAPACITY

	Nov 08	Dec 08	Jan 09	Feb 09	Mar 09	Apr 09	May 09	Jun 09
IT MF Dev Mansfield	-341	42	297	268	305			2
IT Mgmt	2,155	2,515	2,640	2,400	2,64	3,098	2,733	3,296
IT PL GW Dev	-514	480	3,066	3,110	3,163	2,209	1,722	2,269
IT PL MF Dev	871	1,467	2,369	2,225	2,266	176	168	176
IT PL Mgmt	144	168	176	160	176	176	168	176
IT PMS Support	-493	-469	136	160	176	1,136	1,078	1,125
IT Prod Control	6	206	1,136	1,020	1,126	528	504	528
IT Security	432	504	528	480	528	880	840	880
Life Mgmt	720	840	880	800	880	176	168	176
Life PM	144	168	176	160	176	704	672	704
Marketing Manager	156	292	384	640	704	176	168	176
Mktg - Mgmt	144	168	176	160	176	176	168	176
Mktg Comm	144	168	176	160	176	176	168	176

Figure 11.3   The PPM tool is used to provide information needed to make resource allocation decisions. This graphic shows the available capacity of the organization along with the demand for resources from the active projects in the portfolio.

TRAINING

There are two aspects to training. First there's process training and then there's tactical/tool usage training. We recommend that the process training be delivered by someone in-house. The best scenario would be that someone who participated actively in the PPM recommendation committee delivers the process training. This person would be close to the process and perhaps the best person to answer questions that may come up during the training. The PPM tool training, however, is best if delivered by the vendor of the tool. They are best equipped to answer tool-specific questions.

The nature and detail of the training needed will depend on the audience. Different audiences need different levels of training. Some possibilities are:

- Various committee members—To understand the process and their roles and responsibilities
- Resource managers—To be able to use the resource allocation process
- Project managers—To practice project management and estimating
- Contributors—To be able to estimate, track time, resolve issues, create change requests, update status
- New associates and refresher courses—Based on which of the above roles they would fill

CONCLUSION

We covered a number of steps to implement a functional and a sustainable PPM process. Though the approach to implementation may differ based on the size of the organization or the number and/or types of projects in the portfolio, the process fundamentally remains the same. That is, the process has to be designed, appropriate stakeholders have to be engaged, and the process matured by taking a number of baby steps and through a lot of consensus building. At the end of the day, numbers will tell the story of a successful implementation; but it's not to be forgotten that the process supporting those numbers are defined and followed by people. The virtues required for a successful implementation are well known—patience and persistence.

Lesson Learned

Take baby steps. Don't forget the people aspect of the initiative. Attempt to strike an appropriate balance between process and culture.

CHAPTER SUMMARY

Foundational Principles

- PPM is best implemented with a combination top-down, bottom-up approach.
- A strong sponsor is key to PPM implementation success.
- The person heading up the PPM implementation must be a leader.
- The executive chairing the PPM process recommendation committee must be passionate about PPM, understand its value, and have a vested interest in the final outcome.
- Gain stakeholder consensus prior to presenting the PPM process formally to the executive team.
- The process design and implementation should not be perceived as any one individual's crusade.

Foundational Tools

- Project inventory
- PPM implementation communication plan

Lessons Learned

- Define and document all terms—projects, maintenance, system request, and the like—very clearly. Don't assume that everyone has the same definition in mind.
- Gain stakeholder consensus prior to presenting the PPM process formally to the executive team.
- Shared services play a critical role in maturing the PPM process by not allocating their resources to projects unless they're prioritized by the governance committee.
- Be patient in the early stages of the process. Don't get carried away with artificial deadlines and forget the people aspect of the puzzle. Listen to the concerns of the associates and continually tweak the process, not letting it get in the way of productivity.

Chapter 12

Maintaining PPM

By this stage, you have invested lots of time and effort into your people, process, and tools to implement PPM. Great! So, now what? What do you need to do to ensure that this investment continues to pay back the promised value?

Like any new program that has been implemented, the ongoing success of PPM in your organization will be determined by the quality and amount of support and maintenance effort. This is critical to ensure that the principles and processes are not only adhered to but also improved over time. In this chapter we'll discuss the key aspects of what needs to happen for Project Portfolio Management to be sustained after the implementation phase is complete.

DASHBOARDS AND METRICS: THE VISUALS

When most PPM efforts are established, part of the sales pitch to the business leadership usually includes a vision of how it will be presented with wonderful charts and graphs that will clearly depict the organization's performance against goals and objectives, and give a rock-solid view of the state of investments across the business. Metrics can be the most anticipated and least-defined area of PPM, usually because it's very hard to know in advance what a leadership team will actually want, and because it takes time for the data to start flowing. Most PPM teams spend countless hours defining data elements, project scorecards, reports, and process and then stumble when it comes to consolidating and analyzing the data for management review.

Foundational Tool

Project portfolio dashboard

The key to developing metrics starts with an understanding of the maturity of the organization. Many organizations are likely to have been starved of metrics and dashboards due to little or no available data. But now that the data exists, there is a danger that an enthusiastic Project Management Office (PMO) may bombard management with information that it may not be able to absorb. We recommend starting simple and only increasing the complexity and breadth of metrics as the organizational capacity to absorb them grows and matures.

Phase 1: Simple Dashboards

Simple dashboards can be very effective at giving the leadership team a quick and easy-to-understand view of the portfolio. At this stage, we recommend that you focus on charting basic data so that trends can easily be spotted. The following are some simple but effective dashboard metrics:

- *Number of projects active by project phase*. This provides a good sense of the pipeline of projects about to start, underway, or about to deliver.
- *Project dashboard overview*. A high-level view of current active projects. Use Red-Amber-Green (RAG) indicators for various aspects such as whether each project is on budget, running to schedule, and within scope.
- *Spend of active projects against budget*. This is a good visual display for committed budget against total budget within a fiscal year.
- *Resources by project type*. This can provide an excellent view of how resources are spending time or how much time your organization is spending on innovative projects as opposed to sustaining or base business activities.
- *Budget breakdown by major budget transaction class*. This can provide an overview of total annual budget commitment to capital or expense projects.

These five views of project performance will allow an organization to understand the critical fundamentals and provide a solid foundation on which to build greater complexity and detail.

One of the key reasons to start with a relatively simple dashboard is to build management confidence in the aggregated data. As they develop a level of comfort in the information that is being provided by the organization, they will likely demand further detail and analysis to support more effective decision making. It is better to have them demand more information when they are ready than to have them recoil from a mass of what may appear to them at first to be confusing and overly detailed analysis.

Phase 2: Expanding the Dashboard to Include Deeper Analytics

Once you have established the processes and data to create a solid baseline for the dashboard, you are ready to deliver the next phase. This is both an art and a science, and success here will be critical for embedding the PPM function into the fabric of the whole organization. By now the management team is demanding greater analysis, so it's time to get into the details, and this is when the PMO can shine. The key here is to pull together metrics from the information you already have to hand, such as:

- Days between project toll gates. This is a very simple and effective metric that can show compliance levels for review gates and lagging deliverables. It can point out pain points in a process, and highlight areas that need to improve.
- A key metric applicable to all projects is payback, as mentioned earlier. This is a simple view of the benefit versus expenditure in the project.

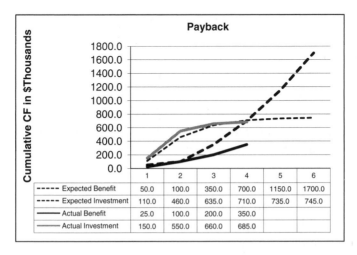

	1	2	3	4	5	6
- - - - Expected Benefit	50.0	100.0	350.0	700.0	1150.0	1700.0
- - - - Expected Investment	110.0	460.0	635.0	710.0	735.0	745.0
—— Actual Benefit	25.0	100.0	200.0	350.0		
~~ Actual Investment	150.0	550.0	660.0	685.0		

Figure 12.1   Payback offers a simple view of the benefit versus expenditure in the project.

Bear in mind a key danger at this stage. Many organizations fall into the data collection spin, where they keep adding to the list of data to collect but never actually get around to analyzing what they have already. A dashboard, and the metrics that go into it, serve one purpose: to provide the business leadership with a solid basis for making decisions. This is where the art part comes in. The successful project portfolio manager knows what the leadership needs to see to make the right decisions. There is no absolute right or wrong answer to what should be in any corporate dashboard. The content should evolve over time as the business becomes more sophisticated—or more mature—in its ability to interpret and react to the data analysis. Get this right and PPM will be seen as a core component of the success of the business; get it wrong and it could end up being ignored.

Foundational Principle

The project portfolio dashboard serves one purpose—to provide leadership with information for making good decisions.

MEETINGS: KEEPING THE PROCESS GOING

Successful, ongoing PPM relies on regular, formal review meetings. It's critical that the business leadership actively supports and participates in these meetings as a visible sign of their ongoing commitment to the investment in time and effort that has already gone into creating it. Everyone in the organization needs to know that the processes they are following and the data they are collecting has to be current, accurate, and complete because it's actively used and relied upon in critical business decisions. A number of different types of review need to be established by the PMO, and each should have a clear structure, purpose, and guidelines. These three are considered essential:

- *Toll gate review.* Setting this up is a key part of implementing the PMO in the organization. However, once the initial enthusiasm has passed, this meeting needs to be established as a critical part of the formal approval process. Key business leaders should attend and vote on allowing projects to proceed through the tollgate or not. The PMO manages the preparation, ensuring that project managers complete the necessary documentation (use of standard templates will make it easier both to prepare the information and to allow participants to become familiar with the information they are being asked to review). During the meeting, the PMO facilitates the discussion and records all issues and decisions.

• *Monthly portfolio reviews.* It is critical for the PMO to get in front of the business leadership on a regular basis to present an overview of the portfolio. This is likely to be the only time when management is exposed to the aggregated data across the entire organization. Since the PMO is the central repository of this data, it is important for them to point out key findings and make recommendations based on their data analysis. Regular reviews will ensure the management team's continued interest in PMO outputs as well as keeping the organization engaged in providing quality data.

• *Process management improvement reviews.* As with any process, there is always room for improvement. The aim of these meetings is to provide those who are key players in the PPM process a forum in which to discuss and recommend improvements and changes to the way things are done. This will give people the opportunity to feel engaged in the process knowing that they have a say in the way it works.

Foundational Principle

Successful PPM relies on regular, formal review meetings.

Many other meetings, checkpoints, and reviews can be set up, but be aware of adding to the administrative burden you may be placing on people.

COMMUNICATION: MAKING SURE EVERYONE IS ON THE SAME PAGE

Effective communication is a key success factor in any endeavor, and maintaining PPM is no exception. This is a critical part of the change management process and will determine how well the organization responds to and embraces the new processes and framework. Communication is a two-way street—to work effectively, it has to be from the top down *and* from the bottom up. Since the PMO sits in the middle, it has a critical role in facilitating this communication flow. There are essentially three levels to these communications:

1. The business leadership delivers communications to support and facilitate the changes in the organization. They can do this in a number of ways:
    - Enable general communications to the organization supporting the changes being implemented, articulating how it will positively affect the organization's—and by extension its peoples'—future.
    - Offer active support of PPM by participation in the tollgate review meetings, demonstrating continuous interest, seeking updates and information, and promulgating and rewarding success stories.
    - Promote the success of PPM to business partners and other external stakeholders by highlighting the continuous improvement in the organization.
2. The Project Management Office (PMO) acts as the communications hub:
    - Communicate all process and procedural best-practices and updates across the organization.
    - Promote and support continual training on processes and tools.
    - Facilitate process review and work with organizational leads to determine process improvements.
    - Capture dashboards and metrics results and provide leadership on interpreting these for decision making.
3. The project management community feeds back successes and lessons learned from overcoming challenges:
    - Understand and implement the processes and tools in everyday practice.
    - Participate in process improvement objectives.
    - Promote processes and tools with business partners and with associates within the organization.

Communications should be coordinated and scheduled to reach across the organization. Accordingly, the PMO should develop a formal communication plan and have it approved for action by the business leadership. This is often forgotten when budgeting for the ongoing role of the PMO, but this critical maintenance activity should be properly resourced to ensure its effectiveness.

MATURITY MODELS: WHERE DO YOU AND YOUR ORGANIZATION STAND?

Once PPM has been implemented, understanding your organization's maturity will help greatly in defining the scope and effort of next steps for improving its effectiveness. There are several maturity models in existence that you can use to assess your company's level of maturity. These can help you to baseline your current performance and establish where you want to be.

Foundational Tool

PPM maturity model

One of the more popular models has five levels of maturity, described as follows (Office of Government Commerce, 2008):

- Level 1: Getting started/awareness/initial
- Level 2: Developing/focusing/repeatable/knowledge
- Level 3: Complying/practicing/competence/defined
- Level 4: Sustaining/exploiting/managed/excellence
- Level 5: Advocating/transforming/optimized

The key imperative of a maturity model is to help an organization assess its level in absolute terms and then develop a roadmap to rise to higher levels. Typically, the following areas are the focus of attention:

- Resource allocation (Are people assigned to the right things?).
- Alignment with strategic imperatives (Are we working on the right things?).
- Visibility into project spend (Are we keeping to budget?).
- Applications (Are we managing our current inventory effectively and efficiently?).
- Balanced portfolio (Is the sum of all projects supporting the total business?).

RESOURCE MANAGEMENT: GETTING YOUR ARMS AROUND THE ORGANIZATION

Most organizations start out by trying to get all of the resources in their organization in alignment through tracking time and effort expended on all activities. This approach is usually very cumbersome, and a newly established PMO can quickly become overwhelmed.

The ultimate goal of resource management is to effectively and efficiently understand resource use and capacity constraints across the organization so that potential problems can be anticipated and avoided. We will focus on the maintenance aspects of resource management.

Assuming that you have implemented a successful resource management process, you will have a skills inventory and defined roles across your organization. This information needs to be maintained so it remains up to date and relevant. The key to success—as well as helping to demonstrate the real value-add of an effective PMO—is to use this information to help the organization plan its talent development and people strategy.

Foundational Principle

Successful PPM relies on up-to-date and relevant information about resources.

A typical problem that this can help to avoid is when changes are made to an organization's technology and platform. Often the latest, cutting-edge technology is implemented without proper attention being given to what it will take to train the workforce to run the new systems. What can then happen is that, rather too late in the implementation process, management realizes that they are not equipped to support the new technologies and they may decide to outsource support to a third party. This can easily leave the company in limbo—they still have some critical legacy systems as well as the new technologies, and the organization remains stuck between the two.

This problem can be avoided once organizations understand their resource skill base and link this to the broader strategy across other functions. In this way, for example, the technology roadmap will help to determine the company's people strategy by focusing efforts on recruiting, retaining, and training the right skills that will be relevant for the future. With this information held in the PMO, the business can then make a determination on whether they want to retain certain skill sets or outsource them to more qualified organizations.

Internal programs need to be established such as mentor and coaching programs to help guide individuals into improving their core knowledge and advancing to higher levels of responsibility. With a good understanding of the skills required by the organization, mentors and coaches can guide individuals into areas that are of benefit to the organization. This type of guidance should flow into individual's goals and objectives so they can track the progress and also keep the organization informed of their career goals.

KEEPING UP THE MOMENTUM

Now that PPM is in place, one of the prime responsibilities of the PMO is to maintain the continuous deployment across the organization so they become embedded into "business as usual."

Chapter 11 discussed how executive sponsorship is critical to the ongoing success of the PPM process. The governance board (or PPM board or decision review board), PMO, and the project management standards committee are described in that chapter along with other groups or committees that can help to keep PPM alive and effective. As mentioned above, communications on a monthly or quarterly basis to the organization at large on progress and success stories is also critical to success. A key element of this is keeping the project managers—the primary information providers to the PMO—aware that their informational updates are being reviewed and that they have a voice in the organization and can help to improve the overall process.

Another way to keep the organizations engaged is to publish a PPM roadmap. This shows the nature of the journey that the organization has embarked upon and how PPM will evolve as the organization matures. This should be used more as a guide, not an absolute direction, and needs to be adjusted and modified as the organization itself changes in response to market realities.

Foundational Tool

PPM roadmap

DEDICATING RESOURCES TO RUNNING THE PMO

The PMO is the "engine room" for deploying and maintaining PPM across the organization, and in driving continuous improvement. It not only facilitates steering committees and review boards, it also manages the day-to-day governance to ensure the PPM processes are being adhered to. In the past, the tendency was not to recognize this as a discreet organizational function, and thus the various roles and responsibilities were scattered across several business functions. Not surprisingly, this obscured the necessary focus and clarity of purpose and there was little formal governance and control oversight. This only leads to failure. Without a dedicated organization to govern and manage PPM, aggregate and analyze information, and enforce and improve the processes, the initial successes will quickly become fragmented and eventually die out.

Foundational Principle

Successful PPM requires a dedicated organization to govern and manage the project portfolio.

The PMO should be recognized as a stand-alone organization and be staffed accordingly. Its remit should be to focus on governance, tools, processes, analysis, and reporting in support of the business leadership. The head of the PMO should, ideally, have a direct reporting line to the senior leadership and not be "hidden" within an existing function. Only in this way will the chances of success be maximized.

CHAPTER SUMMARY

Foundational Principles

- The project portfolio dashboard serves one purpose—to provide leadership with information for making good decisions.
- Successful PPM relies on regular, formal review meetings.
- Successful PPM requires a dedicated organization to govern and manage the project portfolio.

Foundational Tools

- Project portfolio dashboard
- PPM maturity model
- PPM roadmap

Part IV

THE STORY: NINE MONTHS LATER

Chapter 13

Bringing It All Together

Looking outside his office window, John Durham took in the brilliant colors of the fall leaves, which were perfectly reflected in the smooth surface of the lake. He smiled and exclaimed, "Life is good!" He said this to no one in particular—it was late on Friday afternoon and he was the last one in the office.

He thought back to that moment one year ago when, out of the blue, his fraternity brother called. That conversation turned out to be a pivotal moment in his life. Not only had John been able to rekindle the old friendship, but it had led to a transformation in his career.

Just like in their senior year at college, Bill had thrown John a great pass—only this time it was on the field of business in the game of Project Portfolio Management. However, more than a decade ago that single football catch had resulted in the winning touchdown for their frat, but there was no way he could say he had won the PPM game—at least not yet.

Over the last twelve months he had learned an amazing amount from Bill, Chris, and the others at the EPMC. Their advice and counsel was definitely moving him down the field toward the prize, though he was still very much in the trenches, continuing the battle each day, and often into the night. But—and he couldn't help smiling again—just like the thrill of a well-played football game, so it was for him now at Acme. What really made him happy in his work was that he could see the strategic value of his role. Or, maybe it was more of a calling. He'd have to test out that idea at the next EPMC meeting.

Without realizing what he was doing, John reached across his desk and pulled out the special notepad that Bill had presented to him several months back. Bill had said it was to commemorate that dinner when he showed up without any notepaper and he had grabbed the nearest napkin to write on. John kept it on his desk to jot down his bright ideas—this time he made a note about whether PPM could be considered a calling. He chuckled as he scribbled—it wasn't like any notepad you could pick up at the stationery store. No, this was a custom-made pad of paper napkins, each emblazoned with the EPMC logo.

Just then his phone chirped to tell him he had a text message. He glanced at the screen and read, "stil @wk? giv me a cll wen uv a chnc. Bill." He hit the speed dial button on the phone.

"Hi, John! What took you so long to call back?" Bill chuckled.

"Oh, you know, I was just jotting down some amazing ideas on a napkin," John replied.

"Ah, yes! I'd forgotten about that special notepad we made for you," laughed Bill.

"So, what's up, Bill?"

"I just got a call to see a client in your neck of the woods. Are you free for dinner next Tuesday evening?"

"I'm always free to dine with my PPM quarterback," exclaimed John, now laughing out loud.

"What are you talking about?" asked Bill. "Have you been nipping out of that 'tune-townie' hip flask again?"

"No, no, nothing like that. You caught me thinking about how much my life here has changed in just over a year thanks to that conversation we had about football and PPM," replied John.

"Yes, that was an amazing coincidence—and it seems more like five years than one." Bill paused.

"Is everything all right on your end?" asked John, a hint of concern creeping into his voice.

"Oh yes, everything's fine . . . but I was just thinking . . . this past week I've been looking at our portfolio of projects alongside the business as a whole."

"And, are things going okay?"

"Sure—there's no looming crisis at the moment," said Bill. "Actually, quite the opposite. I've been noticing that even though we are making great progress with PPM, there's still a whole lot more we can do. Even though we have captured all major projects in the portfolio net, a big part of the business equation is still missing."

"I'm not sure I understand what you mean," said John. He glanced at his watch. It was Friday afternoon and almost time to head home, but he sensed that Bill had something very interesting to say.

"Well, many of the folks here now understand that looking at the big projects from a holistic, strategic point of view makes great sense. Everyone seems to be happy with the results, but I'm having trouble convincing them that we need to include even more in the PPM process."

"What do you mean by 'even more'?" asked John.

"You know how people are always talking about how their maintenance or operational activities take up a lot of their people's time? When we talk about project resource management, it's rare for anyone to have people available to dedicate fully to projects in the portfolio. They usually claim that they're needed for vital operations or maintenance tasks."

"So, what are you getting at?" John asked.

"I guess I would put it under the heading of Total Performance Management. We need to look at the whole business in order to manage truly holistically. We really need to have operations and projects come together in a company portfolio. It occurred to me today that, in reality, all of our budgets are discretionary."

"Hold on a second!" interrupted John. "Now you're striking at some financial principles that I thought were set in stone. You know, fixed costs versus variable costs and the like."

"That's just it," agreed Bill excitedly. "If you look at a business from the perspective of an extended time horizon, those so-called fixed costs become as variable as the rest. That is, we can choose *not* to spend money on what might be classified as a fixed cost. This would incur consequences down the line and those consequences might really be good for the business when we view things from a full time horizon."

"Hold on there, Bill. You're not making a lot of sense."

"Sorry about that. I get excited about these things and sometimes make leaps in my logic without explaining the steps along the way. It really boils down to reaping benefits when we bring projects and operations together. Make them work in concert with each other and prioritize all work, not just the things we call projects today. Which brings me to the concept that maybe, just maybe, everything we do is a project—it's just a matter of scale. The same principles of strategy and prioritization apply to operations when we view all work as simply a series of projects."

"You've certainly piqued my interest with what you're saying, Bill. I sense you're on to something here. When I think back over the past year and how embedded in my mind the five PPM questions, benefits realization, business cases, and all the other principles you shared with me have become, I'm sure you can do the same with Total Performance Management. I have three bulging ring binders on the shelf next to my desk full of materials and ideas we've shared and I still have that original napkin in a sheet protector right in the front of the binder."

"That's great, John!" Bill said with a note of relief in his voice. "I needed to talk to someone about this and didn't mean hold you up on a Friday afternoon."

"That's okay, Bill. I really don't mind. Actually, just as you were sending me a text message I was thinking how PPM has become more of a calling than a job for me here at Acme. There's something about thinking at a level that transcends the day-to-day drudgery that helps to dull the pain of existence—if you know what I mean! And now I may have to go out and get a new binder and label it Total Performance Management."

"You just might, John!" laughed Bill. "In fact, I might have to do the same thing!"

"Okay, see you Tuesday night," said John. "I'll get dinner reservations and pick you up at your usual hotel at seven."

"You got it! See you then, and bring that new binder . . . "

As John hung up he had a feeling that John and the EPMC were going to be as intimately linked to his work over the next twelve months as they had been over the past twelve months.

References

Aristotle (2004). *Politics: A Treatise on Government* (W. Ellis, trans.). Retrieved January 27, 2009, from http://infomotions.com/etexts/gutenberg/dirs/etext04/tgovt10.htm.

Butler Group (September 2005) *Measuring IT Costs and Value Maximising the Effectiveness of IT Investment*. East Yorkshire, England: Butler Group.

Chew, M. M., et al. (Spring 2006). Managers' Role in Implementing Organizational Change. *Journal of Global Business and Technology*, 2(1). Retrieved January 20, 2009, from www.gbata.com/docs/jgbat/v2n1/v2n1p6.pdf.

Cooper, R. G. (2000). Product Innovation and Technology Strategy. *Research Technology Management*, 43(1), 38.

Cooper, R. G. (2005). *Winning at New Products: Pathways to Profitable Innovation*. Retrieved January 20, 2009, from www.stage-gate.com/downloads/Winning_at_New_Products_Pathways_to_Profitable_Innovation.pdf

Cooper, R. G., Edgett, S. J., & Kleinschmidt, E. J. (1997). Portfolio Management in New Product Development: Lessons from the Leaders—I. *Research Technology Management*, 40(5), 16.

Cooper, R. G., Edgett, S. J., & Kleinschmidt, E. J. (2002). *Portfolio Management for New Products*, 2nd ed. Cambridge, Mass.: Perseus Publishing.

Department of Defense (2000). *Clinger-Cohen Act of 1996 and Related Documents*. Washington, D.C.: Department of Defense. Retrieved January 20, 2009, from www.army.mil/armybtkc/docs/CCA-Book-Final.pdf.

Design (2009). In *Merriam-Webster Online Dictionary*. Retrieved January 20, 2009, from www.merriam-webster.com/dictionary/design.

Di Maio, A. (2006). *Government Business Cases and Portfolio Management: The Essential First Steps*. Stamford, Conn.: Gartner.

Jenner, S. (2009). *Realising Benefits from Government ICT Investment: A Fool's Errand?* Kidmore End, UK: Academic Publishing.

Kendall, G., & Rollins, S. (2003). *Advanced Project Portfolio Management and the PMO*. Fort Lauderdale, Fla.: J. Ross Publishing.

Light, M., Rosser, B., & Hayward, S. (4 January 2005). *Realizing the Benefits of Project and Portfolio Management*. Stamford, Conn.: Gartner.

Markowitz, H. (1952). Portfolio Selection. *Journal of Finance*, 7(1), 77–91.

McClure, D. (2007). *What Frequently Derails IT Portfolio Management in Government?* Stamford, Conn.: Gartner.

Office of Government Commerce (2008). Maturity Models. Retrieved January 20, 2009, from www.ogc.gov.uk/tools___techniques_maturity_models.asp.

Politics (2009). In *Merriam-Webster Online Dictionary*. Retrieved January 20, 2009, from www.merriam-webster.com/dictionary/politics.

Project Management Institute (2008). *A Guide to the Project Management Body of Knowledge*, 4th ed. Newtown Square, PA: Project Management Institute.

Reyck, B. D., et al. (February 2005). The Impact of Project Portfolio Management on Information Technology Projects. *International Journal of Project Management*, 23, 524–537.

Saaty, T. L. (1994). How to Make a Decision: The Analytic Hierarchy Process. *nterfaces*, 24(6), 19–43.

Sanwal, A. (2007). *Optimizing Corporate Portfolio Management: Aligning Investment Proposals with Organizational Strategy*. Hoboken, NJ: John Wiley & Sons.

Schmidt, M. J. (2002). *The Business Case Guide*, 2nd ed. Boston: Solution Matrix.

Sharp, P., & Keelin, T. (March 1998). How SmithKline Beecham Makes Better Resource-Allocation Decisions. *Harvard Business Review*.

Stratton, M. J. (2004). *Business Case Development and Analysis*. Paper presented at the 2004 Crystal Ball User Conference, Denver.

Zander, R. S., & Zander, B. (2000). *The Art of Possibility*. Boston: Harvard Business School Press.

Index